D1358695

REMOVED FROM THE
ALVERNO COLLEGE LIBRARY

Changing the Faces of Mathematics

Perspectives on Gender

Series Editor

Walter G. Secada
University of Wisconsin—Madison
Madison, Wisconsin

Editors

Judith E. Jacobs
California State Polytechnic University—Pomona
Pomona, California

Joanne Rossi Becker
San Jose State University
San Jose, California

Gloria F. Gilmer
Math Tech, Inc.
Milwaukee, Wisconsin

510.71
p467

ALVERNO COLLEGE LIBRARY
MILWAUKEE, WI

National Council of Teachers of Mathematics
Reston, Virginia

Copyright © 2001 by
THE NATIONAL COUNCIL OF TEACHERS OF MATHEMATICS, INC.
1906 Association Drive, Reston, VA 20191-9988
(703) 620-9840; (800) 235-7566; www.nctm.org
All rights reserved

Library of Congress Cataloging-in-Publication Data

Changing the faces of mathematics : perspectives on gender / editor Judith Jacobs.
          p. cm. — (Changing the faces of mathematics)
     Includes bibliographical references.
     ISBN 0-87353-496-4
       1. Mathematics—Study and teaching—United States—Social aspects. 2. Mathematical
ability—Sex differences. 3. Women in mathematics—United States.   I. Jacobs, Judith E.,
1943- II. Series.

QA13 .C45 2001
510′.71—dc21

00-054896

The publications of the National Council of Teachers of Mathematics present a
variety of viewpoints. The views expressed or implied in this publication, unless
otherwise noted, should not be interpreted as official positions of the Council.

Printed in the United States of America

Contents

Introduction

Joanne Rossi Becker

Judith E. Jacobs

Why at the end of the twentieth century are we editing a book on gender equity? The "gender gap" in mathematics has disappeared, has it not? Note how gender equity was discussed in the first draft of the new standards proposed by the National Council of Teachers of Mathematics (NCTM, 1998, p. 25):

> Inequity is pervasive, but it can be corrected in mathematics instructional programs. Consider the case of gender equity, which has been a concern in mathematics education for at least several decades. In the 1960s and 1970s, numerous studies and surveys indicated a substantial gap between males and females with respect to mathematics participation and performance. In the past two decades, great progress has been made in addressing this issue.... Although gender inequity still exists in some schools and mathematics classrooms, it is clear that the "gender gap" has been substantially reduced.

The last twenty-five years have brought considerable progress in women's participation in mathematics at the precollege level (Campbell 1995; Silver, Strutchens, and Zawojewski 1997). For example, data from the Sixth National Assessment of Educational Progress show that 62 percent of females and 60 percent of males had taken second-year algebra; 20 percent and 19 percent, respectively, had taken precalculus; and 9 percent and 12 percent, respectively, had taken at least a year of calculus. In fact, this calculus course was the only one in which females continued to be underrepresented in this national sample (Silver, Strutchens, and Zawojewski 1997).

Women's performance on the SAT college entrance exams has produced less positive results than those described for the participation rates (Groves and Cooper 1999). The data in table 1 of SAT scores for 1999 graduates indicate persistent gender differences favoring males across ethnic and racial groups. It is important to note, though, that women of Asian/Pacific Islander background had higher SAT mathematics scores than men from all other ethnic and racial groups except white men. The gender difference was only 7 points. These discrepancies across and within ethnic and racial groups make the analysis of gender differences complicated. One cannot simply refer to African Americans, Latinos, or Asian/Pacific Islanders. A failure to explicitly examine data about girls and women who are African American, Latina, or Asian/Pacific Islanders is sexist. These ethnic and racial groups are no more monolithic than is the female sex.

During the past half century, the percentage of degrees in mathematics that have been earned by women has increased. As illustrated in table 2, from 1949–50 to 1994–95, the proportion of bachelor's and master's degrees earned by women has doubled and the proportion of doctor's degrees has almost quadrupled. Although there is almost gender parity at the undergraduate level, there are still substantial discrepancies in the proportion of women who earned advanced degrees. Of particular concern is the 17 percent decrease in participation and completion rates as women go from the bachelor's to the master's degree and from the master's to the doctor's degree.

In fact, women represent less than 15 percent of employed doctoral scientists and engineers in the following fields (National Science Foundation 1996): computer science, mathematics, agricultural and food science, environmental sciences,

Table 1
SAT Scores of Graduating Seniors, 1999

Race or Ethnicity	Verbal		Math	
	Male	Female	Male	Female
Asian/Pacific Islander	502	495	579	541
African-American	432	435	434	415
American Indian / Alaska Native	486	481	499	467
Hispanic				
Latin, South, or Central American	471	457	488	446
Puerto Rican	462	450	470	433
Mexican or Mexican-American	459	448	476	441
White	531	524	548	512
Other	515	508	537	494
Overall Average	**509**	**502**	**531**	**495**

Source: The College Board, *1999 College-Bound Seniors, National Report*, SAT I Mean Scores and Standard Deviations for Males, Females, and Total by Ethnic Group. Available from www.collegeboard.com.

chemistry, geology, physics and astronomy, economics, and engineering. At this rate of progress, when can we hope for parity in some of these disciplines?

Although some feminists argue that women are making rational choices by not selecting career fields that are male dominated and hierarchical, and have the potential to involve them in work incongruent with female values, such as military applications (Becker 1990; Fennema 1996; Harding 1986), many others condemn this continuing underrepresentation despite twenty years of intervention. Fox and Soller discuss this issue in chapter 1 in this volume.

As research on gender and mathematics has matured, it has evolved from positivist approaches, in which deficit models were often implicit in the framing of the research, to the use of feminist paradigms (Fennema 1996). In a deficit model people who excel in mathematics are identified and their characteristics are contrasted with those of people who do not do well. Interventions designed from such models usually try to change those who do not do well to be more like those who excel. In mathematics, this notion suggests a song sung by Henry Higgins in *My Fair Lady:* "Why Can't a Woman Be More Like a Man?" Even language has changed, with discussion moving away from "sex differences" to "gender differences" in acknowledgment that gender is socially constructed and that the differences are not biologically determined. Many scholars have learned to

Table 2
Percent of Degrees in Mathematics Earned by Women by Level

	Bachelor's	Master's	Doctor's
1949–50	22.6	19.5	5.6
1959–60	27.2	19.1	5.9
1969–70	37.4	29.6	7.8
1979–80	41.5	33.1	13.6
1984–85	46.2	32.9	15.5
1989–90	45.7	38.1	17.8
1994–95	46.8	39.2	22.1

Source: U.S. Department of Education, National Center for Education Statistics. *Digest of Education Statistics 1997*, Table 291. Available from nces.ed.gov/pubs/digest/97.

acknowledge and support, rather than deny, gender differences, while working to ensure that the European American male model is not regarded as the ideal toward which all must strive. Theoretical frameworks such as *Women's Ways of Knowing* (Belenky et al. 1986, 1997) have stimulated new scholarship that promises to furnish more information on teaching and learning mathematics than older paradigms did (Becker 1996 [see other papers in this special issue of the journal]). This work raises questions. Do girls and women bring alternative ways of knowing and alternative concerns into the mathematics classroom? If so, what are the implications for the mathematics education of girls and women?

Women's Ways of Knowing presents a theoretical model of how women "come to know." This model has five perspectives: *silence, received knowing, subjective knowing, procedural knowing* (separate and connected), and *constructed knowing*. Becker (1995) and Jacobs (1994) present interpretations of this model in mathematics. In the silence and received knowing perspectives, the teacher is the ultimate authority and the student learns only by listening. In the subjective knowing perspective, an important one in women's development according to Belenky and her colleagues, knowledge derives from within, from what feels right. This perspective legitimizes "women's intuition." Men and women handle this perspective very differently: whereas men assert their right to their opinions, women often feel that it is *just* their opinion. In the procedural knowing perspective, the voice of reason emerges and the learner begins to evaluate the validity of an argument. However, again there seems to be a gender difference in this perspective. Whereas men look to propositional logic, argumentation, and rigor (separate knowing), women are more likely to seek access to other people's knowledge, focusing on intuition, creativity, conjecture, and experience (connected knowing). In the constructed knowing perspective, the learner tries to integrate knowledge that is constructed through experience. The context becomes an important part of the consideration of a problem; for example, what can be proved is dependent on axioms being assumed (Jacobs 1994; Jacobs and Becker 1997).

"Traditional" ways of teaching mathematics, which stress certainty, deduction, logic, argumentation, algorithms, structure, and formality, may be particularly incompatible with the ways in which many women learn (come to know). Researchers have hypothesized that this different learning style might account for women opting out of mathematics and related careers. A number of researchers (Becker 1995; Jacobs and Becker 1997) call for connected teaching of mathematics, which stresses many elements. Among these are developing students' voices, affording opportunity for firsthand experience, listening to students, believing versus doubting, and supporting versus challenging. There is evidence that these approaches can provide both more success in mathematics for young women and more positive attitudes (Becker 1996; Buerk 1996; Morrow 1996). Others have begun to question the discipline of mathematics itself. Is it as value-free as many mathematicians argue? Does the traditional mathematics classroom environment convey false conceptions of what mathematics really is (Buerk 1996)?

THE MATHEMATICS TEACHER'S ROLE IN ENSURING EQUITY

These newer perspectives on gender issues in mathematics have influenced the eclectic chapters included in this volume. A number of the chapters challenge us as teachers to consider our role in ensuring equity in mathematics education for girls and women.

Mau and Leitze's chapter asserts that constructivist teaching has the potential for altering the imbalance of power extant in many mathematics classes. By encouraging all students, male and female, to be actively involved in classroom discourse, the constructivist teacher can provide "voice" to young women who frequently avoid dynamic participation in mathematics classes. Some questions remain for us: Is constructivist teaching necessarily feminist (Damarin 1990)? Is it clear that constructivist teaching automatically obviates the need for an awareness of the potential for gender imbalance in classroom interactions?

Ahlquist calls for more radical change in the teaching of mathematics. In her chapter, she makes the case for critical multicultural mathematics teaching that connects race, ethnicity, gender, and social class. She challenges us as mathematics teachers to design curriculum that acknowledges these issues as equity and social justice issues, and she encourages social activism to eliminate inequities. Critical multicultural mathematics instruction builds on students' interests, learning styles, and cultures, which makes the content relevant to students' lives and useful for addressing social problems. We refer the reader to Kitchen and Lear (2000) for a fully developed example of this teaching approach in practice.

The Wilson and Hart chapter contributes excellent action-research ideas for the mathematics teacher who would like to influence her or his own practice by collecting data related to gender issues in her or his classroom. They briefly synthesize research findings related to gender in achievement, participation in the study of mathematics, teacher-student interaction patterns, and affective variables, then suggest action-research ideas relevant to each topic. Classroom teachers will find a wealth of practical ideas for improving their own teaching.

Mitchell and Calahan examine the interplay between gender and assessment in mathematics. They suggest ways in which teachers might modify their assessment practices to eliminate gender bias.

Ultimately the relationship between an individual and mathematics is personal. Several chapters in this book explore this aspect of gender issues. Two chapters suggest ways in which mathematics can be found in myriad everyday activities. Hancock studied the mathematics used in daily work by four African American seamstresses. Her chapter exemplifies connecting mathematics to the personal world of women as called for in *Women's Ways of Knowing*; most women prefer to learn in a connected manner: "The most trustworthy knowledge comes from personal experience rather than the pronouncements of authority" (Belenky et al. 1986, p. 113). Gilmer's paper investigates the use of mathematics in the personal experiences of African American women. Her fascinating description of how hairstylists plait hair, and the mathematics inherent in both the process and the product, demonstrates the relationship of how both a career choice and one's culture can affect this important type of knowledge. The chapter by Oliver is actually a transcript of an interview with a male mathematics educator who has struggled on his journey to understanding gender equity in mathematics. Oliver candidly relates his feelings as a European American male while participating in in-depth sessions dealing with equity. While recognizing his privileged position in U.S. society, he acknowledges the presence of defensiveness and guilt when others relate painful experiences of discrimination. Oliver's sharing of his feelings reminds us how difficult our personal struggle may be before we can fully understand the complexities of gender equity.

Another personal story is the experience of a Latina mathematician as told by Valdés. She relates her sad experiences as a Latina who excelled in mathematics in school. Placed in an accelerated high school program, Valdés had the same teacher for three years. The lack of encouragement from this teacher, luckily, did not have tragic consequences. But the low expectations for a Latina student are quite telling; how many talented Latinas are being overlooked in our schools today?

Becerra's chapter offers insight into the struggles of two women of color as they attempt to assume leadership in mathematics education, especially on issues of equity. These women have experienced a lack of respect and credibility even in their own school districts. The difficulties inherent in assuming leadership on issues of equity, particularly for children of color, are vibrantly retold through the women's own words. It is a cautionary tale for anyone working on equity issues, yet it is ultimately encouraging because these women have found strong support for their work through equity networks.

Campbell and Campbell-Kibler address parents and students with advice about making decisions at important points in the mathematics education of young women. However, we think that the points made in this mother-daughter chapter are relevant to classroom teachers as well, since many of us are called on to provide counseling and educational advice for our students. We end this collection of chapters written from a personal perspective with a poignant poem about Hypatia by Manville; we recommend that you read it when you are ready to channel your anger into action.

MODEL INTERVENTION APPROACHES

Other chapters offer exemplars of interventions that encourage young women in mathematics. Perham and Pierce's chapter describes a residential program for tenth- and eleventh-grade young women that has two purposes: to increase awareness of interesting careers in science, mathematics, and engineering and to increase skills important to pursuing such careers. McCoy's chapter illustrates how the Internet can enable students to investigate the lives of women mathematicians and scientists.

Buerk and Oaks highlight the importance of mentoring young women and of providing role models for those who are successful in mathematics and science. Through an example from literature, they stress how role models can help young women retain self-confidence and value their strengths, especially in science or mathematics. Although we may not all be able to be active mentors, the Buerk and Oaks's chapter illustrates how anyone can offer role models, albeit fictional, to whom our students can relate.

The chapter by Streitmatter, Blair, and Marasco discusses an intervention that is receiving more attention in the United States in the 1990s: single-sex classes in mathematics (American Association of University Women Educational Foundation 1998). This approach remains controversial. Some critics believe that it violates the 1972 Title IX legislation that mandated coeducational classes. Others have philosophical objections, believing that gender equity will be achieved only when teachers find ways to design classrooms that offer a positive and equitable milieu for both men and women, better reflecting real-world social interactions (Mael 1998). Streitmatter, Blair, and Marasco document the importance of the women-only experience to the student participants in a mathematics and science magnet school. Although the teacher started out ambivalent, she ended the three years as an advocate of the program. The description of the differences in instruction and student experiences is fascinating and thought-provoking reading from which all mathematics teachers can glean important implications for their own teaching.

The Forgasz, Leder, and Lynch chapter offers us a glimpse of related work on gender equity in Australia. The authors report in this chapter on two studies, one of a single-sex setting, which is legal in Australia, the second of university students. The single-sex study differs somewhat from that of Streitmatter and her colleagues; Forgasz and her colleagues found that short-term benefits to the young women were not maintained over a longer period. This chapter includes classroom activities appropriate for high school or college students that raise awareness of gender and other equity issues in mathematics.

Keynes, Olson, Cohen, and Bibelnieks report on the impact of intervention for middle school young women in a program designed for mathematically talented students. After identifying the factors related to lower female enrollment in their program for talented youth, the authors created a set of mathematics enrichment activities intended to increase female persistence in the program.

Finally, Kort's chapter discusses an intervention at an earlier age range: the University of Rochester Math, Science, and Computer Camp for girls aged 8–12. As Kort notes, we have evidence that girls begin to lose interest in mathematics as early as third grade. It is essential, she maintains, that this age group sees careers in mathematics and science as accessible as well as exciting.

COMMON THEMES

Although these chapters offer a varied look at issues of gender and mathematics, we perhaps can garner some commonalities from the collection. Many of the chapters delineate the need for, and efficacy of, instructional methods and curricular approaches that are aspects of feminist pedagogy (Jacobs 1994; Jacobs and Becker 1997). Jacobs and Becker discuss four principles of feminist pedagogy that can build a multicultural, gender-equitable mathematics classroom: using students' own experiences in developing curriculum; incorporating writing into instruction; using cooperative learning; and developing a community of learners. Chapters in this volume illustrate these principles explicitly or implicitly. For example, the Gilmer and Hancock chapters show us mathematics in action—in everyday experiences from which curriculum can be built. Ahlquist's chapter explicitly connects issues of race, ethnicity, gender, and social class in the mathematics curriculum. Mau and Leitze's chapter illustrates the importance of students' voices in the mathematics classroom, voices that can be activated in writing or orally. The intervention strategies discussed by several authors incorporate collaboration in the learning process. Read all the chapters to truly understand the many facets of feminist pedagogy in action.

This provocative collection will challenge you in your perspectives on gender equity. We hope that it will also stimulate you to action, whether to examine your own classroom teaching from a feminist viewpoint, to consider how your curriculum should change to be more inclusive, to enlist in local interventions to encourage women in mathematics, or to engage in research on gender. Only with the involvement of all educators will we change the face of mathematics and ensure the attainment of true gender equity.

REFERENCES

American Association of University Women Educational Foundation. *Separated by Sex: A Critical Look at Single-Sex Education for Girls.* Washington, D.C.: AAUW Educational Foundation, 1998.

Becker, Joanne Rossi. "Graduate Education in the Mathematical Sciences: Factors Influencing Women and Men." In *Gender and Mathematics: An International Perspective,* edited by Leone Burton, pp. 119–30. London: Cassell, 1990.

———. "Women's Ways of Knowing in Mathematics." In *Equity in Mathematics Education: Influences of Feminism and Culture*, edited by Pat Rogers and Gabriele Kaiser, pp. 163–74. London: Falmer Press, 1995.

———. "Research on Gender and Mathematics: One Feminist Perspective." *Focus on Learning Problems in Mathematics* 18 (Winter, Spring, and Summer 1996): 19–25.

Belenky, Mary F., Blythe M. Clinchy, Nancy R. Goldberger, and Jill M. Tarule. *Women's Ways of Knowing: The Development of Self, Voice, and Mind*. 10th anniversary ed. New York: Basic Books, 1986, 1997.

Buerk, Dorothy. "Our Open Ears Can Open Minds: Listening to Women's Metaphors for Mathematics." *Focus on Learning Problems in Mathematics* 18 (Winter, Spring, and Summer 1996): 26–31.

Campbell, Patricia B. "Redefining the Girl Problem in Mathematics." In *New Directions for Equity in Mathematics Education*, edited by Walter G. Secada, Elizabeth Fennema, and Lisa Byrd Adajian, pp. 225–41. New York: Cambridge University Press, 1995.

Damarin, Suzanne K. "Teaching Mathematics: A Feminist Perspective." In *Teaching and Learning Mathematics in the 1990s*, 1990 Yearbook of the National Council of Teachers of Mathematics, edited by Thomas J. Cooney, pp. 144–51. Reston, Va.: National Council of Teachers of Mathematics, 1990.

Fennema, Elizabeth. "Mathematics, Gender, and Research." In *Towards Gender Equity in Mathematics Education: An ICMI Study*, edited by Gila Hanna, pp. 9–26. Dordrecht, Netherlands: Kluwer Academic Publishers, 1996.

Groves, Martha, and Richard Cooper. "Ethnic Gap Widens in SAT College Exam Scores." *Los Angeles Times*, 1 September 1999, A1, A16.

Harding, Sandra. *The Science Question in Feminism*. Ithaca, N.Y.: Cornell University Press, 1986.

Jacobs, Judith E. "Feminist Pedagogy and Mathematics." *ZDM Zentralblatt für Didaktik der Mathematik International Reviews on Mathematical Education* 26 (Feb. 1994.): 12–17.

Jacobs, Judith E., and Joanne Rossi Becker. "Creating a Gender-Equitable Multicultural Classroom Using Feminist Pedagogy." In *Multicultural and Gender Equity in the Mathematics Classroom: The Gift of Diversity*, 1997 Yearbook of the National Council of Teachers of Mathematics, edited by Janet Trentacosta, pp. 107–15. Reston, Va.: National Council of Teachers of Mathematics, 1997.

Kitchen, Richard A., and Janet Lear. "Mathematizing Barbie: Using Measurement as a Means for Girls to Analyze Their Sense of Body Image." In *Changing the Faces of Mathematics: Perspectives on Multiculturalism and Gender Equity*, edited by Walter G. Secada, pp. 67–74. Reston, Va.: National Council of Teachers of Mathematics, 2000.

Mael, Fred A. "Single-Sex and Coeducational Schooling: Relationships to Socioemotional and Academic Development." *Review of Educational Research* 68, no. 2 (1998): 101–29.

Morrow, Charlene. "Women and Mathematics: Avenues of Connection." *Focus on Learning Problems in Mathematics* 18 (Winter, Spring, and Summer 1996): 4–18.

National Council of Teachers of Mathematics. *Principles and Standards for School Mathematics: Discussion Draft*. Reston, Va.: National Council of Teachers of Mathematics, 1998.

National Science Foundation. *Women, Minorities, and Persons with Disabilities in Science and Engineering: 1996*. Arlington, Va.: National Science Foundation, 1996.

Silver, Edward A., Marilyn E. Strutchens, and Judith S. Zawojewski. "NAEP Findings Regarding Race/Ethnicity and Gender: Affective Issues, Mathematics Performance, and Instructional Context." In *Results from the Sixth Mathematics Assessment of the National Assessment of Educational Progress*, edited by Patricia Ann Kenney and Edward A. Silver, pp. 33–60. Reston, Va.: National Council of Teachers of Mathematics, 1997.

U.S. Department of Education, National Center for Education Statistics. *Digest of Education Statistics 1997*. Table 291. Available from nces.ed.gov/pubs/digest/97.

Psychosocial Dimensions of Gender Differences in Mathematics

Lynn H. Fox

Janet F. Soller

In the twenty-eight years since the passage of Title IX in 1972, which prohibited gender discrimination in education, women have entered colleges and the workplace in record numbers. Although they have made progress in terms of entry into mathematics and science careers, women have not achieved parity (Vetter 1994). Over the years, educators have questioned the nature and causes of gender differences in achievement in mathematics and science.

In 1976, a series of reports, commissioned by the former National Institute of Education, the forerunner of the U.S. Department of Education, reviewed the research on gender differences related to mathematics and science achievement. The authors of these reports (Fennema 1977; Fox 1977; Sherman 1977) concluded that gender differences were a result, at least in part, of social and educational influences. This conclusion led to two rounds of federal grant competitions for research on the causes of, and remedies for, these differences (Chipman, Brush, and Wilson 1985). We shall discuss briefly the extent of the gender gap today, focusing on what is known about the psychosocial factors that contribute to the differences in test scores, course taking, and career choices.

ACHIEVEMENT AND PARTICIPATION OUTCOMES

Opinions vary about the current extent of the gender gap in mathematics, science, and technology (Cleary 1992; Linn 1992). This situation is largely due to the variety of outcome measures used in different studies, such as achievement on tests, course taking, earned degrees, and career outcomes.

Achievement on Tests

The tests that have been most closely and consistently scrutinized for gender differences relative to mathematics and science are the National Assessment of Educational Progress (NAEP), which measures knowledge among eight-year-olds, thirteen-year-olds, and seventeen-year-olds; the Scholastic Assessment and Achievement tests for mathematics (SAT-M, Math I, and Math II), and the Graduate Record Examination (GRE). The SAT and GRE are used for college admissions and scholarship and fellowship decisions.

NAEP

From 1978 to 1996, the National Assessment of Educational Progress has reported relatively small gender differences in mathematics (U.S. Department of Education 1995, 1997). The differences, though small, consistently favor boys. The gender gap is greatest for the highest scorers as early as grade 4. For example, in the fourth grade in 1996, 24 percent of the boys but only 19 percent of the girls were at or above the proficient level. The gap narrowed a bit for grade 8 to 25 percent and 23 percent for boys and girls, respectively, and dropped to 18 percent and 14 percent by grade 12. For grades 4 and 12, the percents that reached the highest proficiency level of "advanced" were 3 percent for boys and 1 percent for girls.

9

There is a long history of performance differences by as many as 40 points on the mean in favor of males on the SAT-M. In the academic year 1975–76, over 50 percent of the boys who took the SAT-M scored 500 or better, whereas only 33 percent of the girls did that well. Almost twenty years later, in 1994–95, the gap between males and females on mean scores dropped by about 10 points, and the differences in the percents scoring at or above 500 have decreased from about 17 to 14 percent (College Entrance Examination Board 1995, 1996). Even so, this difference is troubling because it affects admissions and scholarship decisions. Several studies cast doubt on the validity of the SAT mathematics tests for predicting the performance of women in college mathematics classes (Bridgeman and Wendler 1991; Strickler, Rock, and Burton 1993; Wainer and Steinberg 1992).

The differences in performance on the GREs have been sizeable and consistent over the past twenty years (Grandy 1994). Clearly this difference must reflect the differences found in course-taking behavior and college majors. This test, like the SAT-M, is used for admissions and scholarship decisions. Thus, the difference in scores can be costly for women. Whether the GRE has the same predictive validity for women as for men needs to be studied.

Course-Taking Differences

Early studies reported large differences in course taking, with men far more likely to take advanced courses such as calculus (Fennema and Sherman 1977). Happily, the percentages of girls in high school mathematics classes have risen. For example, calculus and AP calculus courses enroll almost equal numbers of boys and girls (U.S. Department of Education 1995), and 54 percent of college-intending seniors in 1996 who had taken four or more years of mathematics were girls (College Entrance Examination Board 1996). A sizeable gap still exists, however, between the numbers of boys and girls taking the AP physics and computer science courses (Stumpf and Stanley 1996).

Earned Degrees and Career Choices

Although the gender gap is closing in some career areas, such as medicine, business, and law, it remains fairly sizeable in highly technical applied fields, such as computer science and engineering (Becker 1990). For women, the percent of bachelor's degrees earned in the natural sciences and engineering rose to 12 percent in 1986 but fell to 9 percent by 1991. Although women make up about half of master's degree students and 40 percent of doctorate recipients, they earn about a third of the master's degrees in science and engineering and only 22 percent of the doctoral degrees. In computer science, women earn around 27 percent of the bachelor's and master's degrees and about 15 percent of the doctoral degrees (Vetter 1994). Although about 46 percent of the workforce, women make up only 22 percent of those employed in mathematics, science, and engineering (Hanson 1996).

What Has Changed

There have been important changes over the past twenty-five years. For example, more women are enrolling in college; the gap in test performance on the NAEP has lessened; and course-taking differences, at least in mathematics, have been dramatically reduced at the high school level. However, some sizeable gender gaps remain. For example, test scores continue to favor males on such gate-keeping examinations as the SAT and the GRE. Gaps still exist for earned degrees and career choices, especially in engineering and computer science.

SOCIAL FACTORS

Although the importance of biological differences as determiners of cognitive differences has been under debate, most educators agree that social and psychological factors are crucial in terms of course-taking behaviors and career choices. Attempts to change achievement outcomes in test scores, course taking, or career choice must focus on the psychosocial dimensions of behavior. Many variables have been studied in relationship to test performance, course taking, and career choices. We discuss them within the four clusters: attitudes, support from significant others, classroom climate, and out-of-school learning.

Attitudes toward Mathematics

Although behavior is influenced by many factors, the general attitudes and values that a person holds are relatively constant and influence behavior. For example, if mathematics course taking is optional, we might assume that students will choose to study it if they either enjoy it or perceive it as necessary for future educational and career reasons. In the ensuing discussion, we broadly define the construct of attitudes to include measures of liking for mathematics, self-confidence as a learner of mathematics, achievement motivation, mathematics anxiety, perceptions of mathematics as a male domain, and usefulness of mathematics. Some problems arise when interpreting this body of work because of the nature of self-report data, the overlap among the variables themselves, and the inability to determine causation from correlation data.

Liking for Mathematics

Research on attitudes as measured by simple statements about liking mathematics finds consistent differences between males and females in secondary and college populations (Hyde, Fennema, and Lamon 1990; Kwiatkowski et al. 1993). According to Jackson (1974), the difference was noticeable at the extremes where very favorable or unfavorable attitudes predicted achievement. Gallagher and DeLisi (1994) found that students who used conventional but unsuccessful problem-solving strategies on the SAT-M were also more likely to dislike mathematics and to have lower levels of self-confidence about mathematics. Measures of attitudes have been found to have a small but significant correlation to measures of achievement (Aiken 1986, 1987). A study using grades instead of test scores found that attitudes were more predictive of grades for girls than for boys (Behr 1973). An extremely negative attitude termed math anxiety has been reported by numerous researchers and is said to affect course-taking behavior for girls more than for boys (Tobias 1976).

Self-Confidence as a Learner of Mathematics

Several studies have found girls to report less self-confidence as learners of mathematics (Chipman and Wilson 1985; Junge and Dretzke 1995; U.S. Department of Education 1995), but girls are more likely than boys to report less self-confidence overall (Hyde et al. 1990). All students seem to show some decline in self-esteem over the school years through high school, but the drop is particularly acute for girls. Unlike that of the boys, the girls' self-esteem continues to drop all the way through college (Arnold 1995; Sadker and Sadker 1994). Some evidence shows that self-confidence influences course-taking behavior (Fox, Tobin, and Brody 1979). In a longitudinal study of students from grades 6 through 12, Tartre and Fennema (1995) found that self-confidence in mathematics did predict achievement for both boys and girls. Alas, even girls identified as mathematically gifted report lower levels of self-confidence than gifted boys do (Fox 1982; Fox, Brody, and Tobin 1985; Fox and Zimmerman 1985).

Achievement Motivation

Some theories of motivation have ignored women. For example, McClelland (1961) studied achievement motivation in the 1960s and dropped women from his study because they did not behave like men. Horner (1972) concluded that women feared success, at least in areas viewed as male domains. Two well-developed areas of inquiry about motivation that speak to gender differences in achievement are attribution theory and learned-helplessness. A meta-analysis of research on attribution theory and gender (Whitley, McHugh, and Frieze 1986) found that women were more likely to attribute both positive and negative outcomes to luck than men did. Men were more likely to attribute their performance to their ability, regardless of outcome. Attributing success to ability correlates positively with plans to take more mathematics in high school, especially for girls (Meyer and Koehler 1990).

The Fennema-Peterson Autonomous Learning Behavior Model (Fennema 1984; Fennema and Peterson 1985; Hyde et al. 1990) suggests that causal attributions do affect persistence in studying mathematics whether short-term, as in the willingness to solve a difficult problem, or long-term, as in the willingness to select higher-level courses. Attempts to train students to attribute outcomes to "effort" rather than to luck or ability have had positive results (Heller and Ziegler 1996; Sprinthall and Scott 1989). Licht and Dweck (1984) contend that girls learn to be "helpless" in mathematics because "helplessness" is expected and reinforced by teachers and peers. For example, one study quoted a girl as saying that she screamed when dissecting a frog because she wanted to be seen as "feminine" by peers even though she had no real aversion to doing the work (Shakeshaft 1995).

Perceptions of Mathematics as a Male Domain

Reasearch suggests that girls are less interested in careers in mathematics or science than boys are because society has labeled mathematics a "male domain," but the results are a bit mixed. Fennema and Sherman (1977) reported that males more than females stereotyped mathematics as masculine but concluded that stereotyping did relate to female achievement and course taking. Brush (1985), however, found that stereotyping mathematics as a male domain did not predict course-enrollment plans or preferences in high school. Klein (1989) found that students were likely to sex-type mathematics and science as early as second grade.

A longitudinal study of students from sixth through twelfth grade reported that boys were more likely to perceive mathematics as a male domain. The variable, however, actually correlated with achievement for girls such that girls who did not perceive mathematics as a male domain in sixth grade were more likely to have high levels of achievement in sixth grade. The predictive power extended to eighth, tenth, and twelfth grades (Tartre and Fennema 1995). Unfortunately, in terms of college majors and the world of work, the view of careers in mathematics and science, especially in applied areas such as computer science and engineering, as being male dominated is accurate.

Perceived Usefulness of Mathematics

Why would boys who do not like mathematics or even fear it take more courses than girls who dislike or fear mathematics? Researchers have suggested that the attitude that most affects course taking is the perception of the usefulness of mathematics (Fennema and Sherman 1977; Sherman and Fennema

1977), even when general liking for mathematics was constant (Fox 1977). Differences in usefulness ratings by gender tend to emerge at grade 7, the same grade at which achievement differences are often first noted. Usefulness has been found to be predictive of course-taking behavior (Fennema 1977) especially for above-average-ability girls (Haven 1971). In a longitudinal study of students from sixth through twelfth grade, Tartre and Fennema (1995) reported no gender difference in ratings of the usefulness of mathematics. Cohen and Kosler (1991), however, surveyed all high school students enrolled in advanced mathematics courses in four public school districts in Texas and found males more likely than females to see mathematics as useful in daily life, more likely to agree with the statement that "math has practical value in earning a living," and more likely to agree strongly that "math is necessary for their intended major or career." Usefulness is closely tied to career interests. If girls think that they will not use mathematics in their career choice, they see it as less useful. Hanson (1996) found no difference between boys and girls on beliefs about the usefulness of mathematics per se but did find that girls in grade 7 were less likely than boys to think that they would need science to get a good job and less likely to aspire to jobs in mathematics or science.

Career Interests

Early work on career interests found gender-related differences as early as kindergarten and first grade (Eccles et al. 1993). Even among mathematically gifted youth, more boys than girls are thinking about careers in science, engineering, or mathematics by grade 7 (Fox, Pasternak, and Peiser 1976; Fox and Tobin 1988; Oppler et al. 1993). Jensen and McMullen (1994) found that gender differences in career interest in mathematics and science were not present among gifted fifth graders but were present among gifted sixth graders. By the end of high school, many talented young women have self-selected out of the science pipeline (Hanson 1996). Women gave various reasons for not selecting mathematics, engineering, or science as career options: fear of resentment in the workplace, concern about conflicts with family life, parental discouragement, or the inability to work part-time in this field (Morgan 1992). Eccles (1994) developed a model for understanding women's educational and occupational choices. Although she identified many factors, clearly one of the most powerful, especially in terms of choosing careers in mathematics, was a student's gender-role schema. Therefore, some girls may not seek out information about a career in engineering because it does not fit with their gender schema even though they may have no idea about what engineers actually do.

Support from Significant Others

Although differences in attitudes about mathematics and science, especially in terms of self-confidence and career interests, persist, they are not as pronounced today as in the past. We can speculate that this difference reflects some changes in the attitudes and behaviors of the "significant others" in children's lives that have contributed to the socialization process. Parents, teachers, and counselors reinforce a child's emerging sense of self and view of gender-appropriate behaviors. They influence behavior either by actively encouraging or discouraging choices and behaviors or by acting in more subtle ways, such as ignoring choices or withholding support. Although today few parents, teachers, and counselors are as blatantly sex-role stereotyped about career options for their daughters and female students as were typical adults twenty-five years ago, they may subtly discourage girls from pursuing some educational and career options.

Parents

Parents are important socializing agents. Older studies found that support and encouragement from parents were crucial for girls who chose elective mathematics courses in high school (Haven 1971). Studies of adult women mathematicians also pointed to the importance of parents. For example, Helson (1971) found that the mathematicians she studied were likely to be the oldest daughter in an all-girl family and to have had a father who treated her more like a son. Gifted girls who were highly motivated to achieve in mathematics reported strong support for acceleration and high educational expectations from both parents, but particularly from the mother. Gifted girls more than their matched male peers seemed to need this support to enable them to carry out any risk-taking behavior, such as acceleration in mathematics (Fox 1982; Fox, Brody, and Tobin 1985).

There is some evidence that today's parents are less likely to stereotype mathematical achievement as masculine. In a study of mothers' perceptions about gender differences in achievement, two-thirds of the mothers in the United States sample rated boys and girls equal in mathematics ability. The remaining one-third were evenly divided between seeing girls as better or seeing boys as better (Lummis and Stevenson 1990). Yee and Eccles (1988) found subtle differences in parental perceptions; mothers were likely to attribute a son's success in mathematics to ability, but a daughter's success to effort. Current research concludes that parental influences are still a factor in terms of gifted students' selection of science or mathematics courses in optional summer enrichment programs (Olszewski-Kubilius and Yasmoto 1995).

Some parents are proactive in telling daughters "to be all they can be," but many parents still stereotype mathematics, science, and computers as masculine when they choose toys as presents (Graham and Birns 1979; Lewis 1987). Girls with traditional mothers were less likely to have nontraditional career choices than girls whose mothers were described as "career oriented" (Hoffman 1989). An analysis of longitudinal data for girls who persisted in the science pipeline found that the most persistent girls had mothers who were employed in professions but who had been at home full time while the daughter was in elementary school (Hanson 1996).

Teachers

Severas studies have found teacher encouragement to be important. For example, Luchins and Luchins (1980) cited its importance for women mathematicians. In a study by Solano (1977), teachers reported a more negative stereotype of mathematically gifted girls than of boys. Casserly (1975) found that the teachers who had twice the national percentage of girls enrolled in advanced placement classes in calculus showed few signs of sex-role stereotyping, actively recruited students to their classes, and expected high levels of performance from all their students. Few teachers are intentionally sexist these days. Unfortunately, some do communicate different expectations to boys than to girls and thus unwittingly create a biased classroom (Sadker and Sadker 1994).

Counselors

Numerous studies in the 1970s reported that guidance counselors were very sex-role stereotyped in the advice they gave girls about course selection (Casserly 1975; Haven 1971; Luchins and Luchins 1980). A later study by Hittner and Jacobs (1986) asked counselors to make course-selection recommendations for hypothetical students; no gender bias in recommendations was found. The extent to which counselors have eliminated gender bias in course- and career-guidance activities has not been fully investigated in its natural settings. The facts that more young women are going on to higher education and that there is a reduction of gender differences in course taking suggest that the situation has improved. It is also possible, however, that many schools and school systems have increased their requirements for graduation in terms of the number of mathematics and science courses, thus reducing the influence of counselors on these decisions.

Access to Role Models

Teachers and parents are prime role models. In high school, far more males than females are teachers of mathematics or physics. Although many women with children are working, girls are still more likely to have a father than a mother who is employed as a computer scientist or an engineer. Career interventions that bring women scientists, mathematicians, and engineers to talk with high school students have rarely been subjected to rigorous evaluations. Presumably they do make some difference. There is evidence that extended experiences are effective. For example, in a longitudinal study of gifted girls, those girls who attended a three-week career class were more likely to take mathematics in college or major in mathematics or a related field than were gifted girls who participated in an accelerated summer mathematics program or a control group of girls who had no special summer experience (Fox and Tobin 1988).

Classroom Climate

Classroom climate is a broad term encompassing many dimensions. The discussion here focuses on four areas that have been researched with respect to gender: the absence of women in curricular materials, interactions between students and teachers, gender segregation, and instructional strategies.

Curriculum

Until fairly recently sexist steotypes were the rule in the textbooks. Mathematics textbooks were likely to portray boys as active and girls as passive (Federbush 1974). Rogers (1975) noted that when people were shown in textbooks grouped by gender, boys were grouped by activities such as playing football or baseball but girls were grouped by appearance, such as the length of their hair. Nor were contributions of female mathematicians reflected in most textbooks. A study of elementary school mathematics textbooks over time, however, found that the word problems and examples in the books were relatively free of bias against girls starting in the early 1980s (Nibbelink, Stockdale, and Mangru 1986). Science textbooks, however, continue to show bias. In a review of five secondary science texts, no women scientists were shown and two-thirds of all drawings were of males (Bazler and Simonies 1990). Feldman (1997) also notes that sexist story problems have not been purged from teaching-methods books for preservice teacher training in elementary school mathematics.

Teacher-Pupil Interaction

Studies of teacher-pupil interaction found that teachers spend more time interacting with males and that the quality of interaction is higher for males. Teachers gave males more feedback and encouragement at every grade level and in every subject (Sadker and Sadker 1994). In one study of elementary and secondary mathematics classes, boys interacted more with teachers in every category of interaction—and not because of more disciplinary incidents with boys (Leder 1990). Teacher preparation has not focused enough on making teachers more sensitive to the effects of this subtle form of sexism. Interestingly, a study of gifted students found that the boys were more aware of the discrimination against the girls than the girls were themselves (Feldhusen and Willard-Holt 1993). Perhaps girls notice that they get less attention from teachers but accept the situation as normal.

Gender Segregation

Teachers frequently segregate the girls in their classes and teach to the "male" side of the room (Fox 1996; Sadker and Sadker 1994). The gap in course taking seems to be closing in the high school years, in part because school systems require more courses for graduation for all students, which eliminates some choice. When course taking is optional, a danger exists that advanced or enrichment courses will be gender segregated, especially in mathematics, computer science, and physics. For example, a study of summer enrichment programs for gifted students reported that mathematics and science courses were chosen more often by males (Stocking and Goldstein 1992). As noted earlier, there are still large gender gaps in terms of college majors in applied mathematics and science, such as engineering and computer science, which creates almost segregated classrooms within coeducational schools. Although some educators are advocates of single-sex classes or schools, more research evidence is needed to understand this issue (Riordan 1994). Meanwhile, the "unintentional" segregation that is observed in many educational settings seems to make women "invisible"—not a good situation for girls (Sadker and Sadker 1994). Some advocate separate classes or schools for boys and girls, especially in the middle school years. More research is needed to determine when and how intentional gender segregation is effective.

Instructional Strategies

Traditional views of the classroom focused on teachers as transmitters of knowledge and often ignored individual differences in learning preferences or styles. Some researchers have argued that this has promoted an analytical model of instruction, especially in mathematics and science classes, that has favored males more than females (Belenky et al. 1986; Caffarella 1986; Caffarella and Olson 1993; Collard and Stalker 1991; Gallos 1992). For example, girls tend to like working collaboratively more than competitively. Calls for educational reform and standards, such as those developed by the National Council of Teachers of Mathematics (NCTM), emphasize the importance of the learner's role as a constructor of knowledge, the applications of information and skills to real-world problem-solving situations, and the importance of the social context of learning. These factors translate into more hands-on activities, more cooperative-learning activities, and more emphasis on practical applications (NCTM 1989). Research suggests that such approaches should work better for girls than more traditional teaching methods without any negative consequences for boys (Sadker and Sadker 1994; Slavin and Madden 1989; Tisdell 1993). In a study of fourth graders in mathematics lessons, girls achieved better in cooperative activities than in competitive ones (Peterson and Fennema 1985). Low-achieving boys, however, were not successful in cooperative activities. The authors suggested that teachers should use activities that were both competitive and cooperative, as in Teams-Games-Tournaments, or use activities that were neither competitive nor cooperative. Other researchers have noted that cooperative-learning activities may not benefit girls if the teacher gives no thought either to gender balance or to the instructions given to groups to ensure equal participation (Dillow, Flack, and Peterman 1994). In a study of science learning in high school, students, especially girls, did better with more hands-on science; unfortunately, however, teachers tended not to use this approach in the more advanced courses (Burkam, Lee, and Smerdon 1997).

Out-of-School Learning

One difference between most boys and girls is their use of leisure time. When asked to describe their after-school activities, gifted boys were far more likely than gifted girls to report playing games on the computer (Fox 1982). More recent work suggests that this trend has continued. In a study of several longitudinal cohorts, Hanson (1996) reported that gender differences in access to calculators and computers favored boys. Numerous studies found differences in leisure-time activities along sex-role stereotyped patterns, with boys doing more active things, such as playing sports or video games or building models. Girls were more often found reading, playing with dolls, or participating in domestic activities (Burns and Homel 1989; Johnson 1987). Parents assigned family chores along gender lines. Girls stayed at home to babysit and clean, whereas boys were given more freedom and independence by being asked to run errands (Whiting and Edwards 1988). Although female participation rates in science fairs increased in 1995 as compared with those in 1975, girls were still less likely than boys to do projects in mathematics, physical science, or earth science (Greenfield 1995).

The Influence of Social Factors

Early work on gender differences in achievement in mathematics often sought to identify either biological or social factors that might explain differences in learning outcomes. Social variables that have been consistently found to correlate with either course taking or test performance are attitudes; self-confidence as learners; attribution dimensions in motivational theory; career relevance of mathematics; access to role models; and support from significant others, especially parents. Some have argued that this approach implied a deficit model that looked for ways to make girls more like boys (Fennema and Hart 1994; Jacobs 1994). More recently, the focus of research has shifted to consider such variables as classroom climate and styles of pedagogy as factors that influence learning. Thus, girls may like mathematics less than boys do, not because they dislike the content but because the way it was taught in the past did not emphasize applications or involve group processes in problem solving. Recent efforts to reform the teaching of mathematics endorse more hands-on approaches, thematic teaching, and cooperative learning. All, it is hoped, will result in higher levels of achievement for all students, but especially for girls. Clearly, more research is needed to understand how different learning strategies and activities promote different outcomes for different students.

SOME INTERVENTIONS

Efforts to understand and close the gender gap in science, mathematics, and technology have been numerous, but not always well documented. Early intervention efforts generally fell into one of four categories: efforts to influence girls' attitudes and behaviors in terms of course taking and career choices, efforts to change attitudes and behaviors of significant others, efforts to eliminate biased practices in the classroom, and efforts to change the image of mathematics and science as a male domain (Fox 1980). Currently, the U.S. Department of Education is developing a dissemination system to share information about exemplary and promising programs and practices in a variety of areas, including gender equity in mathematics, science, and technology. A sample of some intervention programs and research on interventions follows.

Interventions Targeting Girls Directly

Self-Confidence

Long and Pagni (1993) described Project MISS, the Mathematics Intensive Summer Session, which specifically targeted college-intending underrepresented minority females who might not consider pursuing higher levels of mathematics achievement. The ongoing success of the programs for the girls came not only from the daily six hours of classwork but also from the evening workshops geared to improving self-esteem and presenting career-development options for women in science and mathematics. In a postprogram survey, 91 percent of the girls in the first summer's program reported gaining confidence in their mathematical problem-solving abilities, and 88 percent reported gaining confidence in their peer problem-solving and relationship abilities. These high gains were replicated in the subsequent summer programs with new groups of girls (Long and Pagni 1993).

Motivation and Attribution

Several studies have shown that students can change their attributions of success and failure from external to internal sources. In a study by Heller and Ziegler (1996), both performance and self-confidence in mathematics improved for girls in junior high school and for female college students who participated in attributional retraining. In another study, eleventh-grade girls fulfilled their community-service requirement by tutoring fourth- and fifth-grade girls. The elementary school girls who were tutored made some academic gains and showed a significant change in their attributions of failure (Sprinthall and Scott 1989).

Access to Role Models

In a three-week summer career class at the end of middle school, gifted girls learned about the uses of mathematics in architecture, pubic health fields, and psychology. The program was extremely effective, and the girls took more mathematics in high school and college and majored in more mathematical or scientific fields than matched control girls (Fox and Tobin 1988). One-day intensive workshops, which are more common approaches to career awareness, have not had such extensive follow-up and evaluation. One program called "Expanding Your Horizons" did find that 90 percent of the students who participated described the conference as helpful toward clarifying their occupational plans, 50 percent said that it helped them realize that they needed more mathematics and science in their high school education, and 85 percent said that it made them more aware of the importance of mathematics and science (Cronkite and Perl 1982).

An Intervention Targeting Significant Others

"Multiplying Options, Subtracting Bias" is a program that focuses on increasing teacher, counselor, family, and peer encouragement for mathematics and science course taking in high school. Videotapes and a facilitator's guide for a two-hour workshop deliver information about the career and educational relevance of mathematics, explain gender differences, and suggest activities to effect change. An evaluation of the program reported mixed results in terms of influencing the attitudes of teachers, counselors, and parents, but it was effective in changing the attitudes of students and the amount of mathematics they intended to study (Fennema et al. 1981).

An Intervention Targeting Classroom Climate

Designed as a yearlong in-service program for teachers, EQUALS focuses on changing the attitudes of mathematics teachers to increase girls' participation in mathematics courses beyond those required for graduation (Kreinberg 1989; Kreinberg and Stage 1981). Teachers are helped to develop teaching materials and strategies that increase girls' self-confidence and competence in mathematics and to incorporate information about careers into the teaching of mathematics. Evaluations of EQUALS concluded that it is an effective program for developing unbiased school environments that facilitate girls' study of mathematics (McCormick and Wolf 1993).

Interventions Targeting Instructional Strategies

An accelerated summer mathematics program for gifted seventh-grade girls that used cooperative-learning strategies effectively promoted girls to accelerate their course taking in high school. Comparable results were not found for girls matched on ability who had no such program (Fox, Brody, and Tobin 1985). With this method, the girls learned a full year of first-year algebra in six weeks of three-hour days, roughly half the hours needed in a regular school year. An experimental class in geometry designed specifically to reduce the gender differences in testing was successful because of the instructional strategies of problem solving related to real-life solutions and a concrete spiral curriculum review (Flores 1990).

Conclusions about Interventions

Many efforts have tried to increase girls' achievement in mathematics and science over the past twenty years. Some have targeted social factors, such as self-confidence as learners, attributional retraining, access to role models, and support from significant others. Others have tried to influence classroom climate and instructional strategies, some directly and others through teacher-education efforts. Although they have reported some degree of success in one form or another, they have not all used the same outcome measures, and only a few have assessed the impact of interventions longitudinally. Thus, although we can be optimistic about the possibility of changing some behavioral outcomes through social manipulation, we cannot specify which models or methods are the most effective or robust.

Research efforts are sorely needed to analyze the interventions more systematically in terms of academic and cost effectiveness. For example, if we can make large gains by changing the way we teach mathematics, investments into changing preservice teacher education may be more effective and less expensive than conducting special programs for girls.

CONCLUSIONS AND IMPLICATIONS FOR PRACTICE

Although we appear to have made progress over the past twenty-five years in closing the gender gap in mathematics, we still recognize some important concerns, especially in terms of college majors and career choices in applied areas related to mathematics, such as engineering and computer science. Clearly, women and girls have fewer overt barriers to education and careers today than they did in 1972. Unfortunately, subtle bias still exists, particularly in classroom climate and access to role models and mentors. Research on the psychosocial dimensions of gender differences in mathematics offers insights into the causes of gender differences in course taking and career choices and suggests strategies for change.

Although girls are taking more mathematics courses in high school than in the past, the attitudinal barriers to career choices in science, mathematics, and technology are already in place by the end of elementary school and increase over time. Thus, interventions for girls are needed at all levels, from upper elementary school through high school. Efforts to increase women's levels of achievement and participation must also target the socializing agents, particularly teachers and parents. Attitudes and behaviors of teachers and parents are important, but the instructional strategies and classroom climate may be crucial factors for female achievement. Fortunately, the strategies that promote higher achievement in science and mathematics for girls—active, hands-on learning and cooperative learning—also benefit boys. Restructuring the mathematics and science classroom to be more gender fair should be the foremost priority in efforts to raise achievement levels and to increase the size and quality of the pool of future scientists and mathematicians.

REFERENCES

Aiken, Lewis, Jr. "Sex Differences in Mathematical Ability: A Review of the Literature." *Educational Research Quarterly* 10, no. 4 (1986–1987): 25–35.

Arnold, Karen. *Lives of Promise.* San Francisco: Jossey-Bass Publishers, 1995.

Bazler, Judith, and Doris Simonies. "Are Women Out of the Picture?" *Science Teacher* 57 (December 1990): 24–26.

Becker, Joanne Rossi. "Graduate Education in the Mathematical Sciences: Factors Influencing Women and Men." In *Gender and Mathematics*, edited by Leone Burton, pp. 119-30. London: Cassell Editing Limited, 1990.

Behr, Anthony N. "Achievement Aptitude in Mathematics." *Two-Year College Mathematics Journal* 4, no. 2 (1973): 72–74.

Belenky, Mary Field, Blythe McVicker Clinchy, Nancy Rule Goldberger, and Jill Tarule. *Women's Ways of Knowing: The Development of Self, Voice, and Mind.* New York: Basic Books, 1986.

Bridgeman, Brent, and Cathy Wendler. "Gender Differences in Predictors of College Mathematics Performance and in College Mathematics Course Grades." *Journal of Educational Psychology* 83 (January 1991): 275–84.

Brush, Lorelei R. "Cognitive and Affective Determinants of Course Preferences and Plans." In *Women and Mathematics: Balancing the Equation*, edited by Susan F. Chipman, Lorelei R. Brush, and Donna M. Wilson, pp. 123–50. Hillsdale, N.J.: Lawerence Erlbaum Associates, 1985.

Burkham, David T., Valerie E. Lee, and Becky A. Smerdon. "Gender and Science Learning Early in High School: Subject Matter and Laboratory Experiences." *American Educational Research Journal* 34 (Summer 1997): 297–332.

Burns, Alissa, and Ross Homel. "Gender Divisions of Tasks by Parents and Their Children." *Psychology of Women Quarterly* 13 (March 1989): 113–25.

Caffarella, Rosemary S. *Psychosocial Development of Women: Linkages to Teaching and Leadership in Adult Education* (Information series no. 350). Columbus, Ohio: ERIC Clearinghouse on Adult, Career, and Vocational Education, 1986. (ERIC Document Reproduction Service No. ED 354 486)

Caffarella, Rosemary S., and Sandra K. Olson. "Psychosocial Development of Women: A Critical Review of the Literature." *Adult Education Quarterly* 43 (Spring 1993): 125–51.

Casserly, Patricia L. 1975. *An Assessment of Factors Affecting Female Participation in Advanced Placement Programs in Mathematics, Chemistry and Physics.* Report to the National Science Foundation, NSF Grant No. GY-11325.

Chipman, Susan F., Lorelei R. Brush, and Donna M. Wilson, eds. *Women and Mathematics: Balancing the Equation.* Hillsdale, N.J.: Lawrence Erlbaum Associates, 1985.

Chipman, Susan F., and Donna M. Wilson. "Understanding Mathematics Course Enrollment and Mathematics Achievement: A Synthesis of the Research." In *Women and Mathematics: Balancing the Equation*, edited by Susan F. Chipman, Lorelei R. Brush, and Donna M. Wilson, pp. 275–328. Hillsdale, N.J.: Lawrence Erlbaum Associates, 1985.

Cleary, T. Anne. "Gender Differences in Aptitude and Achievement Test Scores." In *Sex Equity in Educational Opportunity, Achievement, and Testing*, edited by Joanne Pfeiderer, pp. 51–90. Princeton, N.J.: Educational Testing Service, 1992.

Cohen, Rosetta Marantz, and Joseph Kostler. *Gender Equity in High School Math: A Study of Female Participation and Achievement.* Washington, D.C.: U.S. Department of Education, Educational Resources Information Center, 1991. (ERIC Document Reproduction Service No. 345935).

Collard, Susan, and Joyce Stalker. "Women's Trouble: Women, Gender, and the Learning-Environment." *New Directions for Adult and Continuing Education* 50 (Summer 1991): 71–81.

College Entrance Examination Board. *College Bound Seniors: 1995 Ethnic/Gender Profiles.* Princeton, N.J.: Educational Testing Service, 1995.

———. *College Bound Seniors: 1996 Profile of College Bound Seniors.* Princeton, N.J.: Educational Testing Service, 1996

Cronkite, Ruth, and Teri Hoch Perl. "A Short-Term Intervention Program: Math Science Conferences." In *Women and Minorities in Science: Strategies for Increasing Participation,* edited by Sheila Humphreys, pp. 65–87. Boulder, Colo.: Westview Press, 1982.

Dillow, Karen, Marilyn Flack, and Francine Peterman. "Cooperative Learning and the Achievement of Female Students." *Middle School Journal* 26 (November 1994): 48–51.

Eccles, Jacquelynne S. "Understanding Women's Educational and Occupational Choices: Applying the Eccles et al. Model of Achievement-Related Choices." *Psychology of Women Quarterly* 18 (December 1994): 585–609.

Eccles, Jacquelynne S., Allan Wigfield, Rena D. Harold, and Phyliss Blumenfeld. "Age and Gender Differences in Children's Self- and Task Perceptions during Elementary School." *Child Development* 64 (June 1993): 830–47.

Federbush, Marsha. "The Sex Problem of School Math Books." In *And Jill Came Tumbling: Sexism in American Education* edited by Judith Stacey, Susan Berrean, and Joan Daniels, pp. 178–84. New York: Dell, 1974

Feldhausen, John, and Colleen Willard-Holt. "Gender Difference in Classroom Interventions and Career Aspirations of Gifted Students." *Contemporary Educational Psychology* 18 (July 1993): 355–62.

Feldman, Larry M. "Gender Bias in Mathematics." Paper presented at the annual meeting of the Association for Mathematics Teacher Educators, Washington, D.C., February 1997.

Fennema, Elizabeth. "Influence of Selected Cognitive, Affective and Educational Variables on Sex Related Differences in Mathematics Learning and Studying." In *Women and Mathematics: Research Perspectives for Change* (N.I.E. Papers in Education and Work: No. 8) Washington, D.C.: U.S. Department of Health, Education, and Welfare, 1977.

———. "Girls, Women, and Mathematics." In *Women and Education: Equity or Equality?* edited by Elizabeth Fennema and Jane Ayer, pp. 137–64. Berkeley, Calif.: McCutchan Publishing Corp., 1984.

Fennema, Elizabeth, and Laurie H. Hart. "Gender and the *JRME.*" *Journal for Research in Mathematics Education* 25 (1994): 648–59.

Fennema, Elizabeth, and Penelope L. Peterson. "Autonomous Learning Behavior: A Possible Explanation of Sex-Related Differences in Mathematics." *Educational Studies in Mathematics* 16 (August 1985): 309–11.

Fennema, Elizabeth, and Julia Sherman. "Sex Related Differences in Mathematics Achievement, Spatial Visualization, and Affective Factors." *American Educational Research Journal* 14 (Summer 1977): 51–71.

Fennema, Elizabeth, Patricia L. Wolleat, Joan D. Pedro, and Ann D. Becker. "Increasing Women's Participation in Mathematics: An Intervention Study." *Journal for Research in Mathematics Education* 12 (January 1981): 3–14.

Flores, Penelope V. "How Dick and Jane Perform Differently in Geometry: Test Results on Reasoning, Visualization, Transformation, Applications, and Coordinates." Paper presented at American Educational Research Association, Boston, Mass., April 1990. (ERIC Document Reproduction Service No. ED 320 915)

Fox, Lynn H. "The Effects of Sex Role Socialization on Mathematics Participation and Achievement." In *Women and Mathematics: Research Perspectives for Change.* Washington, D.C.: National Institute of Education, 1977.

———. *The Problem of Women and Mathematics: A Report to the Ford Foundation.* New York: Ford Foundation, 1980.

———. *The Study of Social Processes That Inhibit or Enhance the Development of Competence and Interest in Mathematics Among Highly Able Young Women.* Final Report to the National Institute of Education. Baltimore, Md.: Johns Hopkins University, 1982. (ERIC Document Reproduction Service No. ED 222 037)

———. "Gender and the Self-Fulfilling Prophecy." In *The Self-Fulfilling Prophecy: A Practical Guide to Its Use in Education,* edited by Robert T. Tauber, pp. 136–138. Westport, Conn.: Praeger, 1996.

Fox, Lynn H., Linda Brody, and Dianne Tobin. "The Impact of Early Intervention Programs Upon Course-Taking and Attitudes in High School." In *Women and Mathematics: Balancing the Equation*, edited by Susan F. Chipman, Lorelei R. Brush, and Donna M. Wilson, pp. 249–74. Hillsdale, N.J.: Lawrence Erlbaum Associates, 1985.

Fox, Lynn H., Sara R. Pasternak, and Nancy S. Peiser. "Career-Related Interests of Adolescent Boys and Girls." In *Intellectual Talent: Research and Development* edited by Daniel P. Keating, pp. 242–61. Baltimore: Johns Hopkins University Press, 1985.

Fox, Lynn H., and Dianne Tobin. "Broadening Career Horizons for Gifted Girls." *Gifted Child Quarterly* 11 (January–February 1988): 9–13.

Fox, Lynn H., Dianne Tobin, and Linda Brody. "Sex-Role Socialization and Achievement in Mathematics." In *Sex-Related Differences in Cognitive Functioning: Developmental Issues*, edited by Michele Andrisin Wittig and Anne C. Petersen, pp. 303–34. New York: Academic Press, 1979.

Fox, Lynn H. and Wendy Zimmerman. "Gifted Women." In *Psychology of Gifted Children*, edited by Joan Freeman, pp. 219–43 London: John Wiley & Sons, 1985.

Gallagher, Ann M., and Richard DeLisi. "Gender Differences in Scholastic Aptitude Test—Mathematics Problem Solving among High-Ability Students." *Journal of Educational Psychology* 86 (1994): 204–11.

Gallos, Joan V. "Educating Women and Men in the 21st Century: Gender Diversity, Leadership Opportunities." *Journal of Continuing Higher Education* 40 (Winter 1992): 2–8.

Graham, M. F., and Beverly B. Birns. "Where Are the Women Geniuses? Up the Down Escalator." In *Becoming Female: Perspectives on Development*, edited by C. B. Koop and M. Kirkpatrick, pp. 291–312. New York: Plenum Press, 1979.

Grandy, Jerilee. *Gender and Ethnic Differences among Science and Engineering Majors: Experiences, Achievements and Expectations.* GRE Research Report # 92-03R. Princeton, N.J.: Educational Testing Service, 1994.

Greenfield, Teresa Arambula. "An Exploration of Gender Participation Patterns in Science Competitions." *Journal of Research in Science Teaching* 32 (September 1995): 735–48.

Hanson, Sandra L. *Lost Talent: Women in the Sciences.* Philadelphia: Temple University Press, 1996.

Haven, Elizabeth W. "Selected Community School, Teacher, and Personal Factors Associated with Girls Electing to Take Advanced Academic Mathematics Courses in High School." Doctoral diss., University of Pennsylvania, 1971. Abstract in *Dissertation Abstracts International* 32 (October 1971): 1747.

Heller, Kurt A., and Albert Ziegler. "Gender Differences in Mathematics and the Sciences: Can Attributional Retraining Improve the Performance of Gifted Females?" *Gifted Child Quarterly* 40 (Fall 1996): 200–11.

Helson, Ravenna. "Women Mathematicians and the Creative Personality." *Journal of Consulting and Clinical Psychology* 36 (1971): 210–20.

Hittner, Amy, and Judith Jacobs. "Mathematics versus Science: High School Counselor's Perceptions." *School Science and Mathematics* 86 (November 1986): 559–66.

Hoffman, Lois Wladis. "Effects of Maternal Employment in the Two-Parent Family." *American Psychologist* 44 (February 1989): 283–92.

Horner, Matina S. "Toward an Understanding of Achievement-Related Conflicts in Women." *Journal of Social Issues* 28 (1972): 157–75.

Hyde, Janet Shibley, Elizabeth Fennema, and Susan J. Lamon. "Gender Differences in Mathematics Performance: A Meta-Analysis." *Psychological Bulletin* 107 (March 1990): 139–55.

Hyde, Janet Shibley, Elizabeth Fennema, Marilyn Ryan, Laurie A. Frost, and Carolyn Hopp. "Gender Comparisons of Mathematics Attitudes and Affect: A Meta-Analysis." *Psychology of Women Quarterly* 14 (September 1990): 299–324.

Jackson, Roderick Earle. "The Attitudes of Disadvantaged Students toward Mathematics." Ph.D. diss., Indiana University, 1973. Abstract in *Dissertations Abstracts International* 34 (1974): 3690A.

Jacobs, Judith E. "Feminist Pedagogy and Mathematics." *ZDM: Zentralblatt für Didaktik der Mathematik International Reviews on Mathematical Education* 26, no. 1 (February 1994): 12–17.

Jensen, Rita A., and David McMullen. "A Study of Gender Differences in the Math and Science Career Interests of Gifted Fifth and Sixth Graders." Paper presented at the annual meeting of the American Educational Research Association, New Orleans, April 1994. (ERIC Document Reproduction No. ED379 811)

Johnson, Sandra. "Early-Developed Sex Differences in Science and Mathematics in the United Kingdom." *Journal of Early Adolescence* 7, no. 1 (1987): 21–23.

Junge, Michael E., and Beverly J. Dretzke. "Mathematical Self-Efficacy Gender Differences in Gifted/Talented Adolescents." *Gifted and Talented Quarterly* 39 (Winter 1995): 22–28.

Klein, Carol A. "What Research Says … about Girls in Science." *Science and Children* 27 (October 1989): 28–31.

Kreinberg, Nancy. "The Practice of Equity." *Peabody Journal of Education* 66 (Winter 1989): 127–46.

Kreinberg, Nancy, and Elizabeth K. Stage. "The EQUALS Teacher Education Program." Paper presented at the annual meeting of the American Educational Research Association, Los Angeles, April 1981.

Kwiatkowski, Ericka, Richard Dammer, Jon K. Mills, and Chwan-Shyang Jih. "Gender Differences in Attitudes toward Mathematics among Undergraduate College Students: The Role of Environmental Variables." *Perceptual and Motor Skills* 77 (August 1993): 79–82.

Leder, Gilah. "Gender and Classroom Practice." In *Gender and Mathematics*, edited by Leone Burton, pp. 9–19. London: Cassell Educational, 1990.

Lewis, Michael. "Early Sex Roles Behavior and School Age Adjustments." In *Masculinity/Femininity: Basic Perspectives*, edited by June Machover Reinisch, Leonard A. Rosenblum, and Stephanie A. Sanders, pp. 202–26. New York: Oxford University Press, 1987.

Licht, Barbara G., and Carol S. Dweck. "Determinants of Academic Achievement: The Interaction of Children's Achievement Orientations with Skill Area." *Developmental Psychology* 20 (July 1984): 628–36.

Linn, Marcia C. "Gender Differences in Educational Achievement." In *Sex Equity in Educational Opportunity, Achievement, and Testing*, edited by Joanne Pfeiderer, pp. 11–50. Princeton, N.J.: Educational Testing Service, 1992.

Long, Vena M., and David Pagni. "Targeting Girls: MISS." *Mathematics Teacher* 86 (January 1993): 95–96.

Luchins, Edith H., and Abraham S. Luchins. "Female Mathematicians: A Contemporary Appraisal." In *Women and the Mathematical Mystique*, edited by Lynn H. Fox, Linda Brody, and Dianne Tobin, pp. 7–22. Baltimore: Johns Hopkins University Press, 1980.

Lummis, Max, and Harold W. Stevenson. "Gender Differences in Beliefs and Achievement: A Cross-Cultural Study." *Developmental Psychology* 26 (March 1990): 254–63.

McClelland, David C. *The Achieving Society*. Princeton, N.J.: Van Nostrand, 1961.

McCormick, Megan E., and Joan S. Wolf. "Intervention Programs for Gifted Girls." *Roeper Review* 16 (December 1993): 85–88.

Meyer, Margaret R., and Mary Schatz Koehler. "Internal Influences on Gender Differences in Mathematics." In *Mathematics and Gender*, edited by Elizabeth Fennema and Gilah C. Leder, pp. 60–95. New York: Teachers College Press, 1990.

Morgan, Carolyn. "College Students' Perceptions of Barriers to Women in Science and Engineering." *Youth and Society* 24 (September 1992): 228–36.

National Council of Teachers of Mathematics (NCTM). *Curriculum and Evaluation Standards for School Mathematics*. Reston, Va.: NCTM, 1989.

Nibbelink, William H., Susan R. Stockdale, and Matadial Mangru. "Sex-Role Assignments in Elementary School Mathematics Textbooks." *Arithmetic Teacher* 34 (October 1986): 19–21.

Olszewski-Kubilius, Paula, and Jeff Yasumoto. "Factors Affecting the Academic Choices of Academically Talented Middle School Students." *Journal for the Education of the Gifted* 18 (Spring 1995): 298–318.

Oppler, Scott H., Vicki B. Stocking, David Goldstein, and Laura C. Porter. "Career Interests of Talented Seventh Graders." Paper presented at the annual meeting of the American Educational Research Association, Atlanta, April 1993.

Peterson, Penelope L., and Elizabeth Fennema. "Effective Teaching, Student Engagement in Classroom Activities, and Sex-Related Differences in Learning Mathematics." *American Educational Research Journal* 22 (Fall 1985): 309–35.

Riordan, Cornelius. "The Value of Attending a Women's College: Education, Occupation, and Income Benefits." *Journal of Higher Education* 65 (July/August 1994): 486–510.

Rogers, Margaret Anne. "A Different Look at Word Problems." *Mathematics Teacher* 68 (April 1975): 285–88.

Sadker, Myra, and David Sadker. *Failing at Fairness: How America's Schools Cheat Girls.* New York: Charles Schribner's, 1994.

Shakeshaft, Carolyn. "Reforming Science Education to Include Girls." *Theory into Practice* 34 (Winter 1995): 74–79.

Sherman, Julia. "The Effect of Genetic Factors on Women's Achievement in Mathematics." In *Women and Mathematics: Research Perspectives for Change* (N. I. E. Papers in Education and Work, No. 8). Washington, D.C.: U.S. Department of Health, Education, and Welfare, 1977.

Sherman, Julia, and Elizabeth Fennema. "The Study of Mathematics by High School Girls and Boys: Related Variables." *American Educational Research Journal* 14 (1977): 159–68.

Slavin, Robert E., and Nancy A. Madden. "What Works for Students at Risk: A Research Synthesis." *Educational Leadership* 46 (February 1989): 4–13.

Solano, Cecilia H. "Teacher and Pupil Stereotypes of Gifted Boys and Girls." *Talents and Gifts* 19 (1977): 4.

Sprinthall, Norman A., and Jacqueline R. Scott, "Promoting Psychological Development, Math Achievement and Success Attribution of Female Students Through Deliberate Psychological Education." *Journal of Counseling Psychology* 36 (1989): 440–46.

Stocking, Vicki Bartosik, and David Goldstein. "Course Selection and Performance of Very High Ability Students: Is There a Gender Gap?" Paper presented at the annual meeting of the American Educational Research Association, San Francisco, Calif., April 1992. (ERIC Documentation Reproduction No. ED 372 019)

Strickler, Lawrence J., Donald A. Rock, and Nancy W. Burton. "Sex Differences in Predictions of College Grades from Scholastic Aptitude Test Scores." *Journal of Educational Psychology* 85 (December 1993): 710–18.

Stumpf, Heinrich, and Julian C. Stanley. "Gender-Related Differences on the College Board's Advanced Placement and Achievement Tests, 1982–1992." *Journal of Educational Psychology* 88 (June 1996): 353–64.

Tartre, Lindsay Anne, and Elizabeth Fennema. "Mathematics Achievement and Gender: A Longitudinal Study of Selected Cognitive and Affective Variables (Grades 6–12)." *Educational Studies in Mathematics* 28 (1995): 199–217.

Tisdell, Elizabeth J. "Feminism and Adult Learning: Power Pedagogy and Praxis." *New Directions for Adult and Continuing Education* 57 (1993): 91–103.

Tobias, Sheila. *Overcoming Math Anxiety.* New York: W. W. Norton, 1976.

U.S. Department of Education, National Center for Education Statistics. *The Condition of Education.* Washington, D.C.: Office of Educational Research and Improvement, 1995.

———. *Mathematics Report Card for the Nation and the States.* Washington, D.C.: Office of Educational Research and Improvement, 1997.

Vetter, Betty M. *Status of Women Scientists and Engineers in the United States.* Washington, D.C.: American Association for the Advancement of Science, 1994.

Wainer, Harold, and Linda S. Steinberg. "Sex Differences in Performance on the Mathematics Section of the Scholastic Aptitude Test: A Bidirectional Validity Study." *Harvard Educational Review* 62 (Fall 1992): 323–36.

Whiting, Beatrice, and Carolyn Pope Edwards. "A Cross-Cultural Analysis of Sex Differences in the Behavior of Children Age 3 through 11." In *Childhood Socialization,* edited by Gerald Handel, pp. 271–98. New York: Aldine de Gruyter, 1988.

Whitley, Bernard E., Maureen C. McHugh, and Irene Hanson Frieze. "Assessing the Theoretical Models for Sex Differences in Causal Attribution of Success and Failure." In *The Psychology of Gender: Advances through Meta-Analysis,* edited by Janet Shibley Hyde and Marcia Linn, pp. 102–35. Baltimore: Johns Hopkins University Press, 1986.

Yee, Doris K., and Jacquelynne S. Eccles. "Parents' Perceptions and Attributions for Children's Math Achievement." *Sex Roles* 19 (October/December 1988): 317–33.

Critical Multicultural Mathematics Curriculum

Multiple Connections through the Lenses of Race, Ethnicity, Gender, and Social Class

2

Roberta Ahlquist

"You know, I really just don't give a damn about that multicultural stuff. I mean, I think it's just bull.... Right now I've got more important things to worry about in my teaching."

"Yeah, I agree. I mean, in math I don't really see how you can do it."

I overheard this exchange between two white male prospective high school mathematics teachers as they left my multicultural foundations class at the beginning of the semester. We had just discussed ways to develop critical, antiracist, interdisciplinary, multicultural curricula for all subject areas. Such a reaction is typical among prospective teachers in all fields. Although it is difficult at first to see what multicultural education has to do with mathematics, there are many ways to integrate a critical multicultural approach to the curriculum in any subject area—including mathematics.

THE NECESSITY FOR MULTICULTURAL EDUCATION

Demographics are rapidly changing. By the year 2005, European Americans will be a minority in California; by the year 2050, they will be a minority in the United States. It is fairly clear that within the next ten years, five billion of the six billion people on Earth will be nonwhite. Yet currently in the United States white teachers constitute over 90 percent of the K–12 teaching force, whereas the student population across the nation is increasingly more ethnically, culturally, and linguistically diverse.

The mainstream or dominant culture of the United States has been European American since the genocide of the Native American peoples. The power structure of this country is patriarchal, in great part because of discrimination based on race, gender, and ethnicity. White men continue to maintain power and control over the decisions in our major institutions. A multicultural curriculum is essential if we are to act on the ideals of democracy and social justice for all people, challenge racism and other forms of discrimination, and equitably represent the cultural and language diversity so basic to our nation. In summary, multicultural education is a fundamental means of acting on our commitment to build a truly socially just and democratic society.

What Is White Male Privilege?

Privilege is a special right or immunity enjoyed by a particular person or a restricted group of people. It frees them from certain obligations or liabilities and grants automatic and often unearned advantages, power, or benefits. Not all

25

people who are part of privileged groups are aware of such privileges. They often benefit from their privilege unaware that they are beneficiaries. Often these privileges confer dominance and unwarranted power simply because of a person's ethnicity, gender, or social class.

Privileged people in the United States are generally white, male, middle or upper class, heterosexual, and physically and mentally able. Most members of our society who are born into white male privilege are unconscious of their participation in the reproduction of the status quo, which protects their privileged status and keeps them ignorant of people who are not part of the dominant culture. Indeed, most teachers are not conscious of white male privilege. Most of us are willing to admit that men are more privileged in contrast to women in our society. White upper-class people are extremely advantaged by being born into the European American dominant culture. They receive these advantages primarily on the basis of skin color and social class. Peggy McIntosh (1989) has identified situations in which white-skinned people are advantaged.

- They are not hassled or followed while shopping.
- They see people of their ethnicity widely represented in the media.
- They do not have to educate their children to be aware of systemic racism for their own daily physical protection.
- They can be fairly sure that they have not been singled out because of their ethnicity if they are stopped by a police officer or audited by the Internal Revenue Service.

In contrast, people of color and poor people of all ethnicities are disadvantaged in this society because they are often on the "down" side of power. They cannot automatically assume any of these privileges. More often, they can assume that their race or ethnicity will work against them. They are often suspected, mistreated, underrepresented in the media, discounted compared with white voices, unprotected against systemic or individual acts of racism, isolated, and feared. Teachers need to learn more about how white privilege hurts everyone so that they can play a major role in challenging white privilege in classrooms and society. To combat racism, teachers can then teach students about white privilege. This is an important aspect of multicultural education. Further, white teachers can build alliances across color lines.

What Is Critical Multicultural Education?

A critical multicultural education aims to eradicate institutional and personal discrimination from all aspects of schooling and society. It serves the entire society to have all our students well educated from this perspective. Critical multicultural education opposes any form of gender, race, ethnic, social-class, or language discrimination, in schools and in the larger society. A curriculum that is critical and multicultural helps teachers develop ways to become aware of automatic privileges and to rid ourselves and our classrooms of isms that primarily hurt children of color, girls, poor people, new immigrants, and second-language learners. It sees race, ethnicity, social class, gender, and sexuality as equity and social justice issues. A primary focus is to encourage social activism to eliminate inequities (Nieto 1998; Sleeter 1996).

I use the following brief summary of Sonia Nieto's definition of multicultural education with prospective high school teachers:

1. Multicultural education is antiracist education. It pays attention to all areas in which some students are favored over others in their choice of curriculum materials and resources, in interactions and relationships with students and their communities, and in tracking, assessment, and sorting policies and procedures.

2. Multicultural education is basic education. It is as important as English, mathematics, and science are for living in today's global society.

3. Multicultural education is important for all students—it is about and for everyone, not just for students of color, second-language learners, urban students, or so-called disadvantaged students. All teachers need this perspective: white teachers teaching white students, any teacher teaching students of color, and teachers of color.

4. Multicultural education is global and pervasive. It permeates the school climate; curriculum; physical environment; and relationships among students, teachers, administrators, parents, and the larger community.

5. Multicultural education is education for social justice. Issues of equity and democracy in school and society are implicitly connected. Addressing power relations, economics, and social structures is part of this process.

6. Multicultural education is an ongoing and dynamic process. We continue learning more about multicultural education throughout our lives and changing our perceptions, knowledge, and behaviors.

7. Multicultural education is critical pedagogy. Education is not neutral. It can serve to make us more critical thinkers, better decision makers, more socially just, more democratic in our actions, and more prepared to act on our world in the interests of constructive societal and individual change.

A critical multicultural perspective is controversial because it challenges not only white and male privilege, and the accompanying privileges of the dominant culture, but also the capitalist economic-political system, which creates privileges for some people at the expense of others. Because of the competitive, hierarchical nature of our patriarchal economic system, people of color, poor people, and women have less power and status. If we want to promote equity, social justice, and democracy, as well as cooperation among teachers and students, we need seriously to rethink our curriculum from this perspective.

MATHEMATICS AS A TOOL OF POWER

In our economy, mathematics and science are given more status than music, art, English, or history. As a high-status discipline, mathematics serves as a fundamental selection mechanism to track students into various life directions that can be related or unrelated to the field of mathematics.

Students who are successful in mathematics have opportunities or advantages that other students do not have (Belkhir et al. 1995). Scores on the SAT and other standardized tests determine who goes on to universities, enters higher-paying professions, receives more challenging work, and has some access to move up the social-class ladder. Success in mathematics, therefore, gives certain people cultural capital. If students do not catch on to mathematical concepts quickly, they often fall behind. Those who do poorly in mathematics have lower status, fewer job options, and often lower challenging work.

Who excels in mathematics from race, gender, and social-class perspectives? The critiques about how mathematics reproduces gender inequalities are legion. We cannot, however, work only for gender equity in the teaching of mathematics because race, ethnicity, class, and gender are intrinsically interrelated. Once again, we need to look at power relationships. Who is advantaged and who is disadvantaged in mathematics? How can we diversify and expand the pool of candidates who succeed in this field? How can teachers help broaden access to greater opportunities in the lives of all of their students? How do we design and promote a more inclusive, less mystifying, more exciting curriculum? These are the challenges of a critical multicultural curriculum.

TEACHING MATHEMATICS

Mathematics curriculum has been slow to change over the past 150 years. The content and methodology have remained fairly static. Recently, integrated mathematics has brought some reconsideration to how mathematics should be taught. Research shows that if we were to teach mathematics in a more inclusive, interactive, dialogical, and problem-posing way, we would be more successful with our students (Brenner 1994; Frankenstein 1989; Shor 1992).

Often, the resistance that high school students have toward mathematics discourages prospective mathematics teachers. Far too many high school students are turned off to mathematics because it is too abstract, boring, or dry and unrelated to their daily lives. They resist the way in which abstract concepts are presented in a vacuum; they see little concrete application of the fundamental principles of mathematics. Too often, teachers teach mathematics in a formal, lecture style.

Teachers need to acknowledge that mathematics is far more than numerical and symbolic manipulation. The ways in which mathematics has been typically taught, and the content used to teach mathematics, not only mystify students about the applications of mathematics to peoples' daily lives but also serve to mystify and reproduce the hierarchical socioeconomic system in which we live. Mathematics is not neutral or value-free. The application of mathematics content is political, although it may not seem so (Frankenstein 1987; Powell and Frankenstein 1997). The choices that a person makes in defining the content of mathematics problems reflect a point of view, often that of the dominant white male culture. For example, do mathematics teachers consider the culture and the language skills of their students, especially second-language learners, when teaching specific mathematical concepts and processes?

Not all students learn to think conceptually in the same way. Rote practice, drill, individual worksheets, and more practice do not work well for all students and present a very limited view of the discipline. Mathematics, as a symbolic and syntactical language, has multiple sets of formal rules, which we often assume that all students have learned or picked up through their formal schooling. We cannot assume that all students come to our classrooms with the same constructs. This awareness is the first step in the realization of a more critical multicultural curriculum for high school mathematics teachers.

Finally, to compellingly draw students into the complex and challenging world of mathematics, we must develop culturally relevant content that is connected to the daily life experiences of students. We must demonstrate the application of mathematics as a tool for students to use daily to better understand and improve their lives. The more that mathematics teachers can change abstract concepts into practical applications relevant to their communities of learners, the more meaningful the students' understanding of mathematics will be, as well as the more engaged high school students will become in learning mathematics.

TEACHING MATHEMATICS FROM A CRITICAL MULTICULTURAL PERSPECTIVE

Critical multicultural mathematics instruction is student centered, experiential, hands-on, theme based, and process oriented; it is driven by students' themes, questions, life experiences, and interests. The more we know of our students' origins, ethnicities, learning styles, cultures, and backgrounds, the better we can teach them. This is often a subjective process, involving both mind and heart. Students' and teachers' ideas, questions, and feelings about local and global problems and concepts are considered in the interests of this curriculum. It is collaborative and intellectually challenging and draws on the culture, language, and gender diversity of students.

Here are several approaches to teaching mathematics from a critical multicultural perspective.

Help Students See the Importance of Mathematics

Mathematics teachers can help students develop a broader vision of the importance of mathematics in their lives. Students and teachers need to talk about what kind of mathematics knowledge is important in today's society. Knowledge is power. Just as knowing how to read is power, knowing how to use mathematics is power. Teachers should help students see the need for, and relevancy of, mathematics literacy. Teachers can do this most effectively by posing problems that will help students see the applicability of answers to specific daily-life issues and problems. For example, a teacher might teach scatterplots and simple regression by using data on average beginning salaries in various occupations and the amount of mathematics needed to enter them.

Help Students Understand the Social Context of Mathematics

Teachers need to talk with students about the social and historical contexts of mathematics. How and why were mathematics concepts developed? Whom did they serve historically, and whom do they now serve? Students need to discover the power and status of mathematics in society. For example, mathematics has been used as a selection mechanism to track students into fields that until recently have been accessible to white men far more than to women and people of color (Frankenstein 1989; Sadker and Sadker 1994). How will mathematics be useful in our future? How has mathematics contributed to major discoveries in science, art, and music? What social improvements that affect people's daily lives, society, and world have resulted from the use of mathematics?

Make the Content Global, Interdisiplinary, and Diverse

Rather than rely on dominant-culture texts, draw from curriculum that is ethnically, culturally, and gender relevant. Work to develop interdisciplinary approaches to solving problems. Focus on clarifying the process to solving problems as well as on finding the correct answer. Help students see the material from multiple perspectives. What historical influences have come from Mayan, Aztec, and women mathematicians? What are Incan, Greek, Japanese, Egyptian, and African contributions to mathematics? Use culturally, ethnically, linguistically, and gender-diverse content to compare and contrast approaches to figuring out problems that reflect local or global questions. Create and maintain a network with teachers about new ideas for culturally, linguistically, and gender-diverse lessons.

Make Technology Inclusive

Technology is a tool. Those students who have home access to computers and advanced calculators are at an advantage. All students need to become computer and calculator literate so that they can see that these tools make certain tasks, such as solving systems of equations, easier. The Internet opens up a global perspective on the discipline. Engage students in a discussion about access to, and dependency on, technology and about the need to be able to solve problems when technology is not accessible. Computer software programs, like any other curricular materials, need to be examined to determine whether they are culturally inclusive, nondiscriminatory, and gender sensitive. Interactive mathematics problems on computers, like textbooks, should be deconstructed to ensure that they are culturally and gender inclusive.

Make Mathematics Relevant to Students' Daily Lives

We need to ground mathematics in the culture and context of students' experiences so that students see how mathematics, like any other tool, can help them better understand their world. Mathematics teachers can help students construct appropriate applications to their daily lives. Tap into students' communities and cultures. Figure out how to insert students' experiences with mathematics in their everyday lives into concrete classroom problems. Use their names and their life experiences to construct the curriculum. Instead of assigning mathematics problems in isolation from reality, develop themes and topics in which problems can be posed and solved beyond the classroom, in the surrounding community.

Teachers and students need to look at problems that exist in our communities and society at large. It takes time and effort to define and critique important problems that we need to address in our daily lives. Provide opportunities for students to raise questions and to talk about community issues and injustices they experience in their own lives. Their questions and concerns and the search for answers become the basis for the curriculum.

Raise Awareness about Social Issues

Encourage students to question any aspects of society in which they show some interest. Examining any subject by asking whose interests are at stake, who benefits, and at what cost is especially appropriate for teaching socially and historically grounded, problem-posing mathematics. Raise questions about racism, sexism, social relationships, and power relations. This content can raise awareness about social and economic systems, class and racial issues, and poverty and privilege. It can suggest ways to challenge and change the unequal society in which we live.

Use Mathematics to Address Social Problems

Teach students how to use mathematics as a tool to address and critically assess society and the pressing social issues of the times. Invite them to become researchers of socially relevant problems. We might study with students the origins of poverty, crime, environmental pollution, and the conditions of education and health care in the state and the nation. We might discuss ways to challenge the inequities that exist in the system in which we live. Promote student-generated problems to be solved in the interests of democracy, equity, and social justice.

TEACHING FOR GENDER EQUITY

A female mathematics student-teacher from India described how a mother brought her daughter to her for private mathematics tutoring. The girl's teacher had told the high school student that she did not have to worry about not succeeding in mathematics: "Being a girl, it's OK if you don't do well in math." Nonetheless, the mother brought her daughter to the student-teacher because she wanted the regular mathematics teacher to know that young women are as capable in mathematics as young men. Challenging these views about who is capable of learning mathematics is one way for teachers to support teaching from a bias-free perspective. We need to eliminate the anxieties that some young women feel because of misconceptions about who is and is not capable of learning higher levels of mathematics.

One approach might be to collaborate with students to research mathematics anxiety. How does it reproduce sexism? What are the social implications of having fewer women of color, white women, and men of color in the field of mathematics? How does it affect self-esteem? What are you doing in your class to help students, especially girls, overcome mathematics anxiety? How can parents support teachers in these efforts? Share findings with other mathematics teachers and with parents.

Recently, a female student teacher brought to class a newspaper article that addressed gender biases in the mathematics classroom. It described the lack of sensitivity a white male mathematics teacher showed toward the abilities of young women who were succeeding in mathematics in his classroom. Instead of praising the female student with the highest score on a difficult exam, he used her as the focus of a tirade against the young men in the class. Their sin? They were outscored by a woman. One lesson she may have learned is that girls should not do as well in mathematics as boys. This girl's grades began to drop as a result of this gross act of intimidation, and she never again got the highest mathematics score on an exam. Unfortunately, we can draw on countless stories like this one (AAUW 1991; Orenstein 1994; Sadker and Sadker 1994).

White women and women of color are socialized by the dominant media and popular culture to be sex objects. They are too often viewed as less capable of succeeding in the intellectual challenges expected of males. Critique prime-time media to assess how women, both women of color and white women, are portrayed on television, especially in the ads. Develop graphs and charts to display student research. How many redeeming roles for women did they find in the media? Involve students in the process of acting on their interests. Who benefits when women are portrayed as sex objects or as unequal or inferior to men? At what cost? Discuss the role that statistical information can play in educating others about this social issue. Research how much money is spent on advertising, especially for cosmetics. How do students feel about this? How might that money be better spent?

Encourage student-teachers to write articles for the local paper or mathematics journals, sharing their experiences of teaching mathematics from multiple cultural perspectives. We need more mathematics teachers to see themselves as researchers as they work with students in the classroom. Share questions and ideas with students. Let them know that you believe that everyone can learn mathematics and that all young women should be expected to excel at the same rate as all young men.

GENDER, RACE, AND SOCIAL CLASS EQUITY ARE INTERRELATED

An emphasis on critiquing or highlighting sexism in mathematics is appropriate, if framed in a critical social context, but we need to realize that gender, race, and social class are intricately interwoven. We are socially constructed, and we are not only women of color or white women but also women of a particular social class, ethnicity, and culture. These factors shape who we are and what experiences we have had in life. We need to be sensitive to these complex identities. Why have certain ethnic groups tested lower on standardized mathematics exams? How do we explain this to others and to ourselves? What factors have contributed to African American males doing less well in this field? Women of color face different and multiple oppressions (as people of color and as women). Men of color also are subjected to multiple forms of mistreatment. How can we teach about this in a mathematics classroom?

It is well known that far too many girls in most public schools have been cheated out of an equal mathematics preparation. Girls and boys of color have been cheated many times more, especially if they come from second-language, poor, or working-class backgrounds (AAUW 1995; Orenstein 1994; Sadker and Sadker 1994). Why do boys of color not achieve at the same rate as white boys? A research project in your school would be a useful way to find out whether and why these inequities exist.

Students of color, women, and working-class students are often assumed to be less capable in mathematics. How can we challenge and change this view? Education is the first step. In my teacher-education multicultural foundations class, I share with students my own experiences as a working-class, white high school student with mathematics anxiety. During class one day, as another female student and I were working on a problem at the chalkboard, my white male geometry teacher threw a piece of chalk across the room and said, "Why are girls in here anyway? They can't learn math!" I took this personally and decided that maybe mathematics was really only for men. It took me years to overcome my fears of, and anxiety about, mathematics. Only recently have I come to realize the immense possibilities of knowing more about mathematics and to apply that knowledge to other interests, such as music and art. Think about the different time intervals used in hip hop, jazz, rock and roll, and classical music. I also now realize how mathematics anxiety constrains women's sense of power and control in a gender-unequal and racist economic system. Find out what your students' experiences have been with mathematics and race, class, and gender discrimination. Develop ways to change this pattern of unequal treatment.

One way to educate mathematics teachers about sexism and racism is to unveil and share personal stories of high school students with them. Ask prospective teachers to observe whether students of color and white females have been treated unequally in any of their classes. Engage in class discussions to generate ways to eliminate racial and gender stereotyping and to promote ethnic and gender equity in mathematics. Discuss institutional and individual racism and sexism as a routine part of your curriculum.

If a high school mathematics teacher wants to address institutional sexism, racism, or classism in schools, consider using the vast amount of statistical information about economic inequality, sexism, and racism presented in Peggy Orenstein's *Schoolgirls* (1994) or in Jonathan Kozol's *Savage Inequalities* (1991). Students might research the current differences in funding for urban and suburban schools. Then students could analyze the gross inequalities in funding that are based on ethnicity, race, and social class. Are these gross inequalities grounded in institutional racism, classism, and sexism? How did this occur? What are the effects of land ownership, property taxes, and inequitable schools? Why do rich school districts that spend $15 000 on each student each year exist alongside poor school districts that are able to spend only $2 500 on each student each year? How can this inequality exist after *Brown vs. Board of Education*? This information could be compared on a state-by-state or city-by-city basis, and students could research how state and federal governments could equitably redistribute money to all schools in the United States regardless of the racial and social-class makeup of communities. Until we find solutions to these inequities, we will not eradicate institutional racism, sexism, or other forms of oppression from schooling or society.

TEACHING MATHEMATICS FOR SOCIAL CHANGE

Mathematics applications could be a tool for students to critique and improve the society in which we live. Connecting mathematics to economics helps students learn that economics is often a tool of the people in power and can be used by those who want to exploit others. The economic and social policies of those in power in the United States are often devastating to the poor. Who benefits from major economic shifts, and who is served? Students might be interested in doing an analysis and a critique of the criminal justice system, locally and nationally. Have we moved from a military-industrial complex to a prison-industrial complex? What percentage of crimes are crimes of the economy? What are the implications of cutting school and health-care budgets and of increasing prison and military budgets? These are just a few ways to discuss institutional racism and sexism with statistical documentation.

Student-teachers in my class proposed a research project integrated with social studies. Students would gather statistics that depict ethnic and gender disparities in all aspects of daily life involving the allocation and distribution of land, labor, and capital. This project might give students a broad view of institutionalized racism, sexism, and classism in their county, state, and nation. Students could investigate local and regional housing patterns, land ownership, and government policies that address access to fair housing and jobs, with the goal of developing an analysis of, and alternatives to, current patterns of inequities.

Another idea from these prospective mathematics teachers was to engage students in a critical analysis of racism and sexism in the media. Students in high school watch nearly thirty-five hours of television in a week. That is as much time as they spend in school classrooms. Critically examining the messages that the media send would broaden their understanding of why we see things as we do.

A critical multicultural curriculum for mathematics addresses mathematics content connected to the current daily realities in which we live. For example, my mathematics student-teachers suggested that students could use statistics to learn about the current state of the gender gap when teaching about pay differences between men and women. They also could investigate these questions: How long will it take for women and men to be paid equally for comparable work? What do black women and black men make? What do Latinas and Latinos make? What do Asian American women and men make? What is the basis for differences in wages?

Discuss current social problems with students: the prison-industrial complex that is increasing despite the drop in crime rates; drug arrests; the effects of high student dropout rates on society; and the number of citizens with AIDS. All these data could be graphed and disseminated to students and parents. Students, teachers, and interested parents could organize meetings to share information and ideas for change. A good resource for these lessons is *The New State of the World Atlas* by Michael Kidron and Ronald Segal (1991), which contains maps of the world that are based on social issues and problems.

As teacher educators, we need to situate ourselves and our students in history by providing prospective and experienced teachers with an overview of the socioeconomic context in which mathematics is taught in the United States today. We live in a world in which 85 percent of the world's wealth is owned and controlled by only 4 percent of the population. Who are these 4 percent? We live in a society in which wages for millionaires increased by roughly 240 percent in the 1980s, whereas people earning less than $50 000 in the 1980s (85% of the population) had an average wage increase of just 2 percent. The rich are becoming richer and the poor, poorer.

Rethinking Our Classrooms (Bigelow et al. 1994) is a powerful resource for K–12 teachers who want to make their curriculum more critical, multicultural, and global. It is a collection of articles by teachers for teachers that address equity and social-justice issues from a critical multicultural perspective. Ask students to read and critique some of the articles to assess whether they might want to investigate these topics further. *Rethinking Schools*, the book's corresponding journal, is full of multicultural content that can be used as the basis for discussions of equity and social justice.

How might teachers help students develop a global perspective on mathematics—particularly the historical development of the field? Students would have a more global understanding if they knew the origins of mathematics and understood that not all people in the world approach learning mathematics in the same way. Teachers might include a history of the development of numbers or ask students to research where different theorems originated in the world and their multiple uses and applications over time. This history might include the origins of number systems, including the contributions of mathematicians of color (i.e., Incans, Mayans, Aztecs, Africans, Egyptians) and of women mathematicians from around the world. Use the backgrounds of your students as a basis for a research project on this topic. Contrast and compare Japanese number systems with the United States system or African systems.

SMALL STEPS TOWARD CHANGE ARE POSSIBLE

The prospective high school teachers I described at the beginning of this chapter, who voiced some reticence about developing a critical multicultural mathematics curriculum, did engage in a long dialogue with others in class about the importance of such a curriculum. After completing a foundations class and a multicultural curriculum seminar, both student-teachers developed, taught, and shared two revised curriculum units that were their initial attempts at making their high school mathematics classes more critically multicultural.

One student developed a lesson that required students to graph global population and language statistics. Students had to figure out and graph which five languages were spoken most widely in the world, locate the countries, and define global population percentages of different languages spoken. For his other lesson, he required students to use a concept map to organize students' diverse cultural and gender characteristics, including the country of their origin or that of their ancestors, the country farthest from the United States that they had visited, their favorite ethnic food, and their favorite types of ethnic music, all within the framework of a mathematics curriculum. These are but a few small steps forward in the development of a more critical multicultural mathematics curriculum.

Race, ethnicity, gender, and social class matter. Helping students—and themselves—recognize this fact is a special challenge for socially responsible mathematics teachers. If we truly are to address equity in our demographically changing classrooms and in society in general, we must develop a strong critical multicultural mathematics curriculum.

REFERENCES

American Association of University Women (AAUW). *Shortchanging Girls, Shortchanging America*. Washington, D.C.: AAUW, 1991.

Belkhir, Jean Ait, Maureen Yarnevich, Lawrence Shirley, and Christiane Charlemaine. "Mathematics for All Children: A Multicultural Race, Gender & Class Analysis." *Race, Gender & Class* 3, no. 1 (1995): 125–60.

Bigelow, Bill, Linda Christensen, Stan Karp, Barbara Miner, and Bob Peterson, eds. *Rethinking Our Classrooms*. Milwaukee, Wis.: Rethinking Schools, 1994.

Brenner, Mary E. "A Communication Framework for Mathematics: Exemplary Instruction for Culturally and Linguistically Diverse Students." In *Language and Learning: Educating Linguistically Diverse Students*, edited by Beverly McLeod. Albany, N.Y.: State University of New York Press, 1994.

Frankenstein, Marilyn. "Critical Mathematics Education: An Application of Paulo Freire's Epistemology." In *Freire for the Classroom*, edited by Ira Shor. Portsmouth, N.H.: Heinemann Press, 1987.

―――. *Relearning Mathematics*. London: Free Association Books, 1989.

Kidron, Michael, and Ronald Segal. *The New State of the World Atlas*. 4th ed. New York: Simon & Schuster, 1991.

Kozol, Jonathan. *Savage Inequalities: Children in America's Schools*. New York: Crown, 1991.

McIntosh, Peggy. "White Privilege: Unpacking the Invisible Knapsack." *Peace and Freedom* [Women's International League for Peace and Freedom] (July/August 1989): 10–12.

Nieto, Sonia. *Affirming Diversity*. New York: Longman, 1996.

Orenstein, Peggy. *Schoolgirls*. New York: Doubleday, 1994.

Powell, Arthur, and Marilyn Frankenstein, eds. *Ethnomathematics: Challenging Eurocentrism in Mathematics Education*. New York: State University of New York Press, 1997.

Sadker, Myra, and David Sadker. *Failing at Fairness*. New York: Charles Scribner's Sons, 1994.

Shor, Ira. *Empowering Education*. Chicago: University of Chicago Press, 1992.

Sleeter, Christine. *Multicultural Education as Social Activism*. Albany, N.Y.: State University of New York Press, 1996.

Powerless Gender or Genderless Power?

The Promise of Constructivism for Females in the Mathematics Classroom

3

Sue Tinsley Mau

Annette Ricks Leitze

> And will you try and tell us you've been too long in school?
> That knowledge is not needed? That power does not rule?
>
> Gordon Lightfoot, "Sit Down Young Stranger"

In 1988, we began trying to understand how the educational process affected our lives and those of our students. Compelling social justice issues faced us when we explored curriculum and instruction, especially the way young girls and women were taught mathematics. We found help in understanding this process by examining the relationships of power and knowledge development within the classroom. From the perspectives of two women with advanced degrees in mathematics and in education, a central issue is the power relationship between teacher and student.

Although gender differences in mathematics achievement and course enrollment at the kindergarten through grade 12 levels have decreased, young women continue to avoid mathematics-related college majors and careers (Campbell 1995). A National Research Council (NRC) report states that

> cross-national studies of gender differences in mathematics suggest that most of the differences observed are due to the accumulating effects of sex-role experiences at home, in school, and in society. The gender gap in mathematics widens with increasing exposure to school and society; moreover, in countries with more rigid curricula where mathematics courses are required and students do more homework, gender differences are reduced significantly. (NRC 1989, pp. 22–23)

Few gender differences in the performance of mathematical tasks are found prior to age ten. Just before or just after the onset of adolescence, however, gender differences emerge, and they generally favor boys (Friedman 1989). They occur during the tumultuous time of adolescence when both boys and girls are keenly aware of their sexuality and seek a strong sex-role identification. Meyer and Koehler (1990) point out that to a female adolescent, achievement in mathematics may be perceived by her peers as incompatible with a feminine sex role. Because of a strong desire to accommodate peer pressure, many young women elect to end their mathematical studies. In so doing, they begin the process of becoming powerless in mathematics.

In the past, much study has focused on overt sex bias in attitudes within the mathematics classroom. Researchers have examined sex bias in texts, in remarks made by teachers, and in the influence a teacher's gender has on interactions with students. It appears that the differential mathematics performance of males and females is complicated and cannot be explained by any one of these factors (Koehler 1990). Thus, our attention has turned toward differential-treatment studies.

In differential-treatment studies, researchers examine the teaching environment to determine whether males and females are treated differently. If teachers acknowledge the responses of boys more often than those of girls, they send the message that the girls' responses are not as important as the boys' responses. Moreover, if teachers offer help more frequently to boys than to girls, they send the message that girls are not as important as boys. Girls are set up to become powerless. Martel and Peterat (1988, p. 120) claim that

> in the interactions between teachers and students, girls are silenced; they become spectators, wallpaper flowers, listeners of the boys, who, given more time and attention, form the dominant valued core and command the action of the classroom. This command places boys foremost in the minds of teachers as they plan classes so that teaching in very specific ways is directed to boys.

This differential treatment inherently marginalizes girls and their mathematical development.

POWER RELATIONSHIPS IN DEVELOPING KNOWLEDGE

Although mathematics itself[1] is not necessarily power laden, the mathematical canon found in U.S. schools is saturated with human values and epistemology that derive their power from the sanction of traditionally European American male mathematicians. When we teach so that our students become submissive and demand that they understand mathematics in ways that support the traditional mathematical canon, we limit their development. When we teach our mathematics students to "be quiet and listen," we deprive them of the opportunity to create their own meaning, disempower them, and remove their opportunity to develop autonomy. For young women, this removal of autonomy may be particularly serious.

To understand what it means to be disempowered, we must understand empowerment. An evolving definition of empowerment, given by Rosenman (1980), is the "ability of individuals to control their lives and to affect forces which impede such control" (p. 252). Kreisberg (1989) defines empowerment as "a personal transformation out of silence and submission that is characterized by the development of an authentic voice. It is a social process of self-assertion in one's world, intricately linked to themes of dominance and liberation" (p. 5).

For our purposes, empowerment in mathematics means the ability to articulate meaningful mathematical understanding. To us, "self-assertion in one's world" means that a student not only could participate but *would* participate without hesitation or fear of ridicule. We propose that constructivist learning theories, when coupled with teaching from a constructivist perspective, offer the opportunity for girls and women to develop participatory competence; constructivist teachers will, of necessity, offer girls and women the space to become self-assertive in the mathematical arena. In this way, we believe, the transition from powerless gender to genderless power can begin.

---

1. Our view is that mathematics is the study of patterns and shapes. For us, this means that *humans* make generalizations about their experiences and then form a compelling case that justifies their generalizations. The mathematical structure has nothing inherent in it that places a value on the case made by the *human*. It is the human who chooses to accept or reject the case of another. This is where the issue of power becomes part of mathematics.

A LEARNING SPACE FOR FEMALES

Belenky, Clinchy, Goldberger, and Tarule (1986), in their chapter on education for women, relate two stories told to them by respondents. In the first, a science professor asked students to guess the number of beans in a jar. When the students guessed incorrectly, he told them never to trust their senses; that would lead them astray. In the second, a philosophy professor placed a cube on the desk and asked students how many sides it had. When the students said six sides, the professor asked them how they knew this, when they could not see all the sides. Students gave their explanations, and the professor responded by telling them to trust their senses, which would help them make sense of the world.

Belenky and her colleagues claim that the women in their study needed "confirmation that they could be trusted to know and to learn" (p. 195). They needed to know that they were integral parts of the classroom and that their thinking was of interest to the teacher. Their "connected teacher" starts with what the students know and what new knowledge they gain through the act of learning. They suggest that "connected teachers try to discern the truth inside the students" (p. 223). Noddings (1984) echoes this validation of students as learners when she describes a caring teacher of a student who hates mathematics. This caring teacher begins "as nearly as I can, with the view from his eyes: Mathematics is bleak, jumbled, scary, boring, boring, boring. From that point on, we struggle *together* with it" (pp. 15–16, emphasis added).

The importance of the partnership to the educational process cannot be minimized. It is in the presumption that students and teacher are equal partners that students' feelings, beliefs, attitudes—their very existence—are validated; it is in this presumption of validity as human beings that students and teachers are able to develop the relationship necessary for dynamic learning. This partnership and this presumption of validity will, we believe, facilitate students' developing genderless power in mathematics.

BASIC TENETS OF CONSTRUCTIVISM

Constructivism, put simply, is a theory about knowledge and its development. Cobb (1988) asserts that

> constructivism challenges the assumption that meanings reside in words, actions, and objects independently of an interpreter. Teachers and students are viewed as active meaning-makers who continually give contextually based meanings to each other's words and actions as they interact....From this perspective, mathematical structures are not perceived, intuited, or taken in but are constructed by reflectively abstracting from and reorganizing sensorimotor and conceptual activity. They are inventions of the mind. (P. 88)

This process radically changes the traditional notions of teachers and authority. The teacher is no longer the all-knowing disseminator of knowledge. The "I lecture, you listen" attitude is not present in the constructivist's classroom. The cooperative atmosphere exemplified by "we talk, we listen" allows all students to create mathematical meanings with less teacher intervention. Students are sculptors of their knowledge, and teachers are the polishing agents for that knowledge.

What is it about constructivism that makes it particularly appropriate for women? To answer this question, we must consider what feminist pedagogy is. Gore (1993) focuses her thoughts about feminist pedagogy around issues of authority: authority as power and authority as authorship. The authority of determining what the correct knowledge is, possessing the correct knowledge, having the correct answer—all are instances of authority as power. Teachers who determine how students must answer questions and what questions must be answered limit students' authorship of their own knowledge. For constructivist teachers, issues of authority are central to their pedagogy, and they encourage students to becomes authors of their own knowledge.

In the 1990 Yearbook of the National Council of Teachers of Mathematics, Steffe describes "adaptive mathematics teaching." He begins with a student's words: "You just don't understand!" He continues by describing his view of teaching as a beginning high school teacher, a view that had been shaped by some of the best mathematicians in the country. At that point, he viewed mathematics as a target set of understandings for students to master and to demonstrate in teacher-sanctioned, acceptable ways. That is, Steffe imposed his view of mathematics on his students, eliminating their voices and their authority as authors. Steffe discusses his realization that such an approach is discriminatory and excludes almost all students (p. 44). He then advocates "adaptive mathematics teaching" that "shifts the focus of mathematics teaching from a process of transferring information to students to interactive mathematical communication in a consensual domain of mathematical experience" (p. 45).

Simon (1995) proposes that a teacher learns with students. Although the teacher is responsible for choosing lesson topics and materials, the teacher does not come to the situation omniscient. Rather, the students teach the teacher while the teacher teaches students. This situation, students as teachers, elevates the students to the status of collaborators in the educational process. They are neither below nor above the teacher; they are simply individuals whose value in the classroom is the same as the teacher's.

Steffe and D'Ambrosio (1995) define a constructivist teacher as one who interacts with students. If the teacher interacts solely with male students, the stage is set, yet again, for developing powerless females. However, if the constructivist teacher interacts with all the students, females and males alike, the stage is set for females to develop "self-assertion in [their mathematical] world" and they are empowered.

RETURN TO POWER RELATIONSHIPS

We began by discussing power relationships and the potential that constructivist teaching has for altering that power balance. We believe that when females' voices are equally heard along with the teacher's voice and with the males' voices in the mathematics classroom, females will begin to develop the participatory competence in mathematics we described earlier. Constructivist theories of learning and a constructivist teacher necessitate increased student-teacher and student-student interactions. Those interactions lead to conversations that, in time, will validate young women's places in mathematics classrooms (see Rogers [1990]). We note Fennema and Hart's (1995) comments that girls often lose their voices at the onset of puberty, an occurrence that leads to the termination of their participation in mathematics. We hope that constructivist teachers can bring those voices back to the classroom. Only then will we truly have genderless power in the mathematics classroom.

REFERENCES

Belenky, Mary F., Blythe M. Clinchy, Nancy R. Goldberger, and Jill M. Tarule. *Women's Ways of Knowing: The Development of Self, Voice, and Mind.* New York: Basic Books, 1986.

Campbell, Patricia B. "Redefining the 'Girl Problem in Mathematics.'" In *New Directions for Equity in Mathematics Education*, edited by Walter G. Secada, Elizabeth Fennema, and Lisa Byrd Adajian, pp. 225–41. Cambridge: Cambridge University Press, 1995.

Cobb, Paul. "The Tension between Theories of Learning and Instruction in Mathematics Education." *Educational Psychologist* 23 (1988): 87–103.

Fennema, Elizabeth, and Laurie E. Hart. "Gender and the *JRME*." *Journal for Research in Mathematics Education* 25 (1995): 648–59.

Friedman, Lynn. "Mathematics and the Gender Gap: A Meta-Analysis of Recent Studies on Sex Differences in Mathematical Tasks." *Review of Educational Research* 59 (1989): 185–213.

Gore, Jennifer M. *The Struggle for Pedagogies: Critical and Feminist Discourses as Regimes of Truth.* New York: Routledge, 1993.

Koehler, Mary S. "Teachers, Classrooms and Gender Differences in Mathematics." In *Mathematics and Gender*, edited by Elizabeth Fennema and Gilah C. Leder, pp. 128–48. New York: Teachers College Press, 1990.

Kreisberg, S. "Empowerment and Collaborative and Cooperative Learning: Redefining Power in the Classroom." Paper presented at the annual meeting of the American Educational Research Association, San Francisco, 1989.

Martel, Angeline, and Linda Peterat. "A Hope for Helplessness: Womenness at the Margin in Schools." *Journal of Curriculum Theorizing* 8, no. 1(1988): 103–35.

Meyer, Margaret R., and Mary S. Koehler. "Internal Influences on Gender Differences in Mathematics." In *Mathematics and Gender: Influences on Teachers and Students*, edited by Elizabeth Fennema and Gilah C. Leder, pp. 60–95. New York: Teachers College Press, 1990.

National Research Council. *Everybody Counts: A Report to the Nation on the Future of Mathematics Education.* Washington, D.C.: National Academy Press, 1989.

Noddings, Nel. *Caring: A Feminine Approach to Ethics and Moral Education.* Berkeley, Calif.: University of California Press, 1984.

Rogers, Pat. "Thoughts on Power and Pedagogy." In *Gender and Mathematics: An International Perspective*, edited by Leone Burton, pp. 38–46. Singapore: Cassell, 1990.

Rosenman, Mark. "Empowerment as a Purpose of Education." *Alternative Higher Education: The Journal of Non-Traditional Studies* 4 (1980): 248–59.

Simon, Martin A. "Reconstructing Mathematics Pedagogy from a Constructivist Perspective." *Journal for Research in Mathematics Education* 26 (1995): 114–45.

Steffe, Leslie P. "Adaptive Mathematics Teaching." In *Teaching and Learning Mathematics in the 1990s*, 1990 Yearbook of the National Council of Teachers of Mathematics, edited by Thomas J. Cooney, pp. 41–51. Reston, Va.: National Council of Teachers of Mathematics, 1990.

Steffe, Leslie P., and Beatriz D'Ambrosio. "Toward a Working Model of Constructivist Teaching: A Reaction to Simon." *Journal for Research in Mathematics Education* 26 (1995): 146–59.

Teachers as Researchers

Understanding Gender Issues in Mathematics Education

4

Patricia S. Wilson

Laurie E. Hart

Issues of equity in mathematics are important to teachers, students, administrators, parents, and researchers. Specifically, gender equity has been a long-standing concern in mathematics education. This concern has led to a strong, respected, and ever-evolving body of research. Gender equity in mathematics is important because there is evidence that females have not always achieved at the same level as males in mathematics and have not been encouraged to participate in mathematics to the degree that males have. Although the gaps in mathematics achievement and in participation between males and females may be growing smaller, equity for females and males in mathematics education has still not been achieved.

Although there is a large research base about gender and mathematics, applying the results of this research to particular classroom, school, and community contexts can be difficult. As teachers apply these findings to their classroom practice, it is essential for them to understand the roles and limitations of research. Research can and should influence teacher decision making, but research was not designed to give definitive answers about what to do in particular situations. Each teacher needs to assess his or her own classroom and to make decisions on the basis of the research and his or her knowledge of the community, school, students, and curriculum. Action research offers teachers ways to collect relevant data as part of the instructional process and ways to use data to revise their own practice. In this chapter, we offer action research as a link between research about gender issues in mathematics and equitable classroom practice.

RESEARCH ON GENDER AND MATHEMATICS

Research investigations into gender and mathematics have been a significant tradition in mathematics education for more than twenty years. A thorough overview of research trends in the United States and in other countries was reported by Gilah Leder (1992). From 1978 to 1990, gender-related issues made up about 10 percent of the research reported in the *Journal for Research in Mathematics Education*, the research journal of the National Council of Teachers of Mathematics.

Three features of the research on gender are especially striking. First, differences in participation, performance, and attitude between boys and girls tend to be much smaller than differences within the population of girls or within the population of boys. This fact makes looking at individual girls extremely important and cautions us against applying findings about a population to an individual girl or boy. Second, several studies have found rather small differences that favor boys' learning of mathematics. It is reasonable to question the significance of very small indicators, but in a large body of research, the small differences repeatedly favor boys. The combined effect of small differences could be an important piece of the picture. Third, the research on gender varies widely in terms of methodology, populations, geography, research tasks, and classroom settings. For example, there is evidence that patterns of gender differences in mathematics achievement and attitudes are distinct for different ethnic groups (e.g., Brandon, Newton, and Hammond 1987; Jones 1987; Reyes and Stanic 1988). Therefore, it is not surprising that different studies report seemingly conflicting results. As we continue our research investigations, we may find variables that actually account for or reconcile some of the conflicts. At this point, it is important to realize that differences exist among regions of the country, schools, and teachers—and even between different classes of the same teacher. These caveats are important in using research in the classroom.

Leder (1992) expressed the need for a special approach to gender research:

> Supplementing the more common large-scale studies with in-depth small-sample research should provide further insights into the factors that contribute to differences in mathematics learning within as well as between groups and should lead to more constructive ways of counteracting them. Adoption of research paradigms that allow greater attention to individual differences and context-specific problems is consistent with research thrusts in other areas of mathematics education. (P. 617)

Action research offers opportunities to address Leder's concerns.

ACTION RESEARCH

Action research is part of a continuing cycle that allows a teacher to investigate what is happening in the classroom. Since the process is cyclical, the beginning point for action research is not clear, but the components and value of this type of research are well documented (Hopkins 1985; Lytle and Cochran-Smith 1992). As an action researcher, the teacher (1) identifies an issue of interest related to his or her teaching; (2) proceeds to study the issue by reading and discussing and by identifying salient variables; (3) collects useful data; (4) analyzes the data in terms of instructional decisions; and (5) modifies instruction and proceeds with the cycle by studying the modifications. The entire cycle should be integrated into the instructional process; it should not be considered an independent task. Although it is additional effort, action research enriches and vitalizes instruction for both the teacher and the students (Lankford 1993). The integration is the necessary action component. Action research offers a chance to link the research on gender and mathematics and the classroom.

How to Begin Action Research

Each major section of this chapter has three components: general research findings, rethinking the general findings, and action research ideas.

General Research Findings

We identify four broad areas of gender research related to mathematics education and present data and contexts that help characterize the general findings. This component gives background information and suggests issues and readings.

Rethinking the General Findings

We extend the general findings by sharing studies that challenge the findings, clarify the parameters, and warn against overgeneralization. This component informs the reader about important issues related to gender and mathematics and, equally important, raises questions about the findings and about the implications for instruction. Readers should look for connections among the general findings, individual cases, and their own situations.

Action Research Ideas

We have developed ideas for action research that teachers can easily modify and use as part of their classroom instruction. We designed the action research ideas to help teachers gather data in their own classrooms and to analyze data in a way that will inform them about their individual situations.

It has been widely reported that boys perform better than girls in mathematics. Although there are contradictory findings, studies in the United States and international studies report that boys score higher on standardized achievement tests and perform better on higher-level mathematics. We need to be cautious about drawing conclusions from such general statements and heed two important caveats. First, although there are consistent between-gender differences, the difference is quite small compared with the differences among boys and among girls (Leder 1992). Second, it is important to know that the small differences between genders seem to be decreasing (Fennema 1995).

However, we still see more males than females recognized as extremely high achievers. For example, males are disproportionately represented among students with the top 10 percent of scores on standardized mathematics achievement tests (Leder 1990). Mathematics teams are often all male, and substantially more doctoral degrees in mathematics are granted to men than to women (National Center for Education Statistics 1995). Although some of these examples are influenced by females choosing not to participate in particular activities, it is relevant that girls and women are not as well represented in the groups that are identified by traditional measures as the highest achievers in mathematics.

Rethinking the General Findings

Given that boys score slightly higher than girls on mathematics achievement tests, we find it interesting to ponder research that challenges the general finding. When we look at particular categories of achievement in mathematics, we note that girls excel in some areas and boys in others. Females tend to outperform males on tasks with defined procedures, and boys seem to outperform girls when a strategy for solving the problem is less apparent (e.g., Damarin 1995). In 1988, the National Assessment of Educational Progress (NAEP) found that eleventh-grade boys scored significantly higher than girls in higher-level applications, geometry, and measurement (Dossey et al. 1988).

Over the years, gender differences measured by NAEP have been sporadic and difficult to interpret. We do know that when the difference between male and female students has been significant, the overall scores for males have been higher than the scores for females. In 1988, there was no difference in mathematics achievement at the elementary school level (age 9) as measured by NAEP, but in high school (age 17), the boys outperformed the girls even when they had taken the same courses (Mullis, Owen, and Phillips 1990). In 1992, NAEP measured no gender differences at any grade level, but in 1996, fourth-grade boys scored higher than girls (Reese et al. 1997).

When achievement is measured with grades in mathematics classes, the patterns of gender-related differences are very different from those found with standardized achievement tests. The gender-related differences in mathematics-course grades "almost always favor girls, and these differences are quite consistent across samples of varying selectivity for junior high through university mathematics courses" (Kimball 1989, p. 199). In contrast, the average mathematics scores on the old SAT and the new SAT I: Reasoning Test for males was higher than that for females in every year from 1972 to 1996. Possible scores on the SAT range from 200 to 800, and the differences between the average scores for males and females varied from 35 to 46 points (College Board 1996, Table A). These scores represent a 5- to 7-point difference in raw scores on the sixty-item test. We must be cautious in interpreting these scores because raw scores include penalties for guessing and because more females than males take the SAT (College Board 1997a; College Board 1997b).

These findings suggest that drawing conclusions from achievement data is complicated. Girls seem to achieve in different areas and in different ways than boys do. Girls may place a different value on achieving. We do not know whether changes in format and content of assessment in standardized tests could show a different picture.

Action Research Ideas

Given the complex issues related to achievement and the variation that exists by region and even by classroom, action research affords an opportunity for teachers to learn more about gender differences in mathematics achievement in their own classrooms. To understand how boys and girls differ in their patterns of mathematical achievement, it is important to look at mathematical knowledge by content (e.g., measurement, geometry, algebra) and to look at mathematical processes (e.g., calculation, drawing, proof, tables, reasoning). Are there certain areas or processes in which girls excel or in which they are deficient?

One action research project could investigate the kind of information being assessed on teacher-prepared tests. For example, although a test may integrate calculation, reasoning, and problem solving, the teacher could choose to analyze the test items by separating the items that assess calculation from those that assess problem solving or reasoning. By analyzing responses by content or by process, the teacher may gain a sense of strengths and deficiencies by gender. Males and females may have similar test scores but show different profiles of strengths in areas of calculation, problem solving, or reasoning.

Investigating mathematical achievement at the school level is also informative. Faculties who collect mathematics achievement scores at each grade level in their school and examine standardized-test scores by gender and grade level have an opportunity to look for developmental gender patterns. Do gender differences change from grade to grade? Do the scores show gender differences for particular subsections of tests? If so, do they change from grade to grade? Because gender patterns can differ by ethnicity or social class, teachers may want to look at scores by grade level, gender, ethnicity, and social class.

A third action research investigation could explore the impact of different forms of assessment on both females and males. A teacher could try a variety of assessment strategies and look for gender patterns in students' performance in different modes, such as report writing, journals, tests, projects, and oral presentations. Teachers can also experiment with a variety of test items. For example, a teacher could ask students to explain what is wrong with a solution to a given problem or to give a counterexample to a false statement and provide an explanation.

It is also interesting to investigate what students think is the nature of mathematical achievement. A teacher could ask students to create a mathematics achievement test and then investigate whether the test questions differ by the gender of the author. Students may create better tests after they have taken tests that include different kinds of questions and formats. By encouraging students to include a variety of items, teachers may gain insights into what students—boys and girls—consider important in mathematics.

Achievement and affective issues are often confounded. Action research that investigates what students think it means to be a good mathematics student can offer insights into both the students' mathematical achievement and their feelings about themselves as mathematics students. Have students write a journal entry about what it takes to be a good mathematics student. Use this or a similar prompt: "One morning you wake up and you are an outstanding mathematics student! Describe the kinds of things you know and how you feel." Study their responses by looking for whether students talk about knowing rules and formulas or whether they value understanding and processes. Notice any mention of affective characteristics, such as feeling shy, powerful, unpopular, happy, or successful. Are there differences between successful students and struggling students? Are there gender differences in the responses?

ENROLLMENT ISSUES

Participation in mathematics is determined largely by students' enrollment in mathematics courses in school. Leder (1992) reviewed research about the participation of female and male students in mathematics courses in secondary and postsecondary education. She found that once participation in mathematics courses was no longer compulsory, well-documented and persistent differences in enrollment in mathematics courses occured. She found evidence of higher rates of participation in elective mathematics courses by males than by females among both U.S. and Australian high school students. Further, "the pervasiveness of gender differences in participation in postcompulsory mathematics courses [was] confirmed by Schildkamp-Kundiger's (1982) review of international data on gender and mathematics. Participation differences were found to be most marked for high level, intensive mathematics courses and in applied fields requiring such courses as prerequisites" (Leder 1992, p. 607).

The fourth NAEP mathematics assessment was conducted during the 1985–86 school year. Meyer (1989) analyzed NAEP mathematics course enrollments by gender and found little difference between female and male students in course taking through second-year algebra. However, males were more likely than females to take the elective, fourth-year courses of precalculus and calculus. The sixth NAEP assessment reports similar findings (Silver, Strutchens, and Zawojewski 1997). Meyer (1989) also concluded that males were more likely than females to take college-level mathematics courses and to pursue mathematics-related careers.

Challenges to the General Findings

Gender differences in mathematics participation have decreased during the past twenty-five years. A larger percentage of 17-year-old females took second-year algebra, precalculus, and calculus in 1986 than in 1978, and the gender difference in enrollment declined from 1978 to 1986 (Meyer 1989). Data from the National Center for Education Statistics (NCES) (1995) indicate that since 1970–71, the gender gap between men and women earning bachelor's, master's, and doctoral degrees in mathematics and statistics in the United States has declined steadily (table 4.1). Even so, in the United States many more men than women earn undergraduate or graduate degrees in mathematics. Thus, although the enrollment of females in mathematics at the secondary and postsecondary levels has increased since the 1970s, beyond second-year algebra a higher percentage of males than females still enroll in mathematics courses.

Table 4.1
Degrees in Mathematics and Statistics Conferred by U.S. Institutions of Higher Education

Year	Bachelor's degrees		Master's degrees		Doctoral degrees	
	Men	Women	Men	Women	Men	Women
1970–71	15 498	9 439	4 149	1 546	1 154	95
1980–81	6 614	4 819	2 106	968	656	119
1990–91	8 178	7 132	2 446	1 595	837	199

Action Research Ideas

Perhaps most important in the enrollment picture is that patterns of enrollment vary by region of the country, state, district, school, and teacher. Enrollment data by gender, particularly for ability-grouped classes and for elective mathematics courses, can indicate the level of gender equity in a particular district or school.

Because of the importance of mathematics courses, it is crucial that teachers and school administrators be aware of gender patterns in enrollment in these courses. At the elementary and middle school levels, mathematics is required for all students. However, the tracking of students into ability-grouped mathematics classes is pervasive, particularly in the middle grades. It is important for teachers and administrators to uncover any gender (or race) differences in students' placement into ability groups.

At the high school level, recognizing gender or race differences in enrollment in mathematics courses is important. Are there differences in the percentage of female and male students who enroll in remedial or slow-paced mathematics courses; college-preparatory mathematics courses, including algebra and geometry concepts; and advanced courses, such as precalculus or calculus? If differences in enrollment are found, do they occur for particular courses or teachers or are they schoolwide? Are there consistent patterns of enrollment differences for a high school and its feeder middle schools?

TEACHER-STUDENT INTERACTION ISSUES

Many studies of teacher-student interaction in mathematics classrooms have examined how the experiences of boys and girls differ. The patterns vary from classroom to classroom and from teacher to teacher. Further, both girls and boys vary in their level of participation in classroom interaction. However, evidence from a number of studies suggests that teachers interact less with girls than with boys (Becker 1981; Brophy 1985; Fennema and Peterson 1987; Hart 1989; Sadker and Sadker 1994). For example, in one study of twelve seventh-grade classrooms, Hart (1989) found that teachers had more public interactions with boys than with girls (though these differences were not large). In their study of thirty-six fourth-grade mathematics classes, Fennema and Peterson (1986) found that teachers had significantly more interactions about mathematics with boys than with girls: "Teachers initiated more math contacts, both high level and low level, with boys than with girls" (p. 35). For example, teachers called on boys more frequently than girls for both answers to word problems and explanations about how to solve word problems.

Rethinking the General Findings

In a qualitative case study of one seventh-grade mathematics classroom, Stanic and Hart (Stanic and Reyes 1987) analyzed the teacher's treatment of the seventeen students in the class (five black females, three black males, five white females, and four white males) and the consequences of this treatment for individual students. They cautioned that "findings of race differences and sex differences in mathematics should not be overgeneralized" (Stanic and Hart 1995, p. 258). They found that gender and race were important in understanding the interactions that students had with their teacher, but that neither gender nor race was adequate in categorizing the classroom interaction, achievement, or attitudes of these middle school mathematics students (Stanic and Reyes 1987). They suggested that at the least, "it is crucial to look at the interaction of the categories of race and gender" (Stanic and Hart 1995, p. 258).

These and other findings about teacher-student interactions in mathematics classes paint a complex picture of relationships and interactions between teachers and students. In some situations, differential treatment of students can lead to negative consequences for students in a classroom setting, but in others differential treatment is an appropriate adjustment of instruction to the individual characteristics and needs of students. What is clear, however, is that teachers typically are unaware when they call on one group more than another (Sadker and Sadker 1994).

Action Research Ideas

Aside from the complexity of the patterns of interaction between teachers of mathematics and their female and male students, teachers should be aware of both the number and type of interactions they have with different students in their classes. It is important for teachers to find out which students are most salient in class interaction, which students participate but do not dominate, which students say little or nothing, which students are most or least likely to be called on by the teacher, and which students seek out or avoid interacting with the teacher.

Mathematics teachers will find it worth their time and effort to identify the patterns of teacher-student interactions with individual students in their classes, paying particular attention to patterns by gender and race. One approach to collecting teacher-student interaction data is to videotape a class or a series of classes and then count the number and type of interactions with each student. A similar approach is for two teachers to spend time in each other's classes to observe and record teacher-student interactions.

One excellent observation instrument for recording classroom interaction was described by Stallings, Needels, and Sparks (1987). A peer observes another teacher's classroom and records the number and type of interactions the teacher has with the class as a whole and with individual students. Beforehand, the observed teacher prepares a seating chart that includes students' names along with their gender and race. The coding system is relatively simple and easy to use (fig. 4.1).

After the observation, the teacher who was observed uses the data to analyze his or her interaction patterns. Together, the teachers use the patterns of interaction as they discuss possible strategies for changing any undesirable patterns. Confidentiality is important in this process and should be established as a ground rule before any observations are conducted. It is also important that any peer observations be conducted on variables that are as objective as possible. The data collected should be straightforward records of observable classroom events, which prevents teachers from feeling that they are being judged.

Pairs of teachers who work together in this way can become an important support for each other. Teachers who have the opportunity to observe other classrooms often learn a great deal from the experience.

? = Knowledge-level questions (direct questions): require a right answer, simple recall of facts; include review questions.

      Example A: What is 9×6?

      Example B: What is the sum of the measures of two supplementary angles?

H = Higher cognitive questions: require students to think, apply, interpret, analyze, synthesize, create, or evaluate.

      Example A: What reasoning did you use to solve that problem?

      Example B: Prove: If a trapezoid is isosceles, its opposite angles are supplementary.

U = Checks for understanding: require students to show that they understand the content or procedures of the lesson; call for summarizing, explaining, comparing. Students may respond orally, in writing, or by using hand signals, flash cards, or slates. Checks for understanding or comprehension can be given to the total group or to individuals.

      Example A: What is wrong with problem 6?

      Example B: Explain in your own words what an isosceles triangle is.

+ = Praise or acknowledgment: students' academic responses, actions, or products are praised or acknowledged.

      Example A: Thank you for your contribution. That helps solve the problem.

      Example B: Great job!

C = Correction: students' academic responses or products are wrong, and the teacher corrects them.

      Example A: Yakeesha, tell John the correct answer.

      Example B: No, Mary, it's an equilateral triangle.

G = Guided correction: students' academic response is wrong or incomplete, and the teacher guides, probes, restates.

      Example A: Let me ask the question in another way.

      Example B: You're almost right. Let's go through the steps again.

S = Social comments: teacher makes a social comment to a student. Even though stated as a question, a social comment is coded.

      Example A: I like your haircut.

      Example B: So, how was the soccer match yesterday?

– = Reprimand: teacher reprimands behavior. This code always refers to behavior.

      Example A: Class, you will all be on silent lunch today.

      Example B: Tim and Mary, please stop talking now!

---

* = Student initiates: students initiate remarks or questions to the teacher. These are all coded *. Be sure to code the teacher's response if there is one.

      Example A: Chris asks, "What is our assignment for tomorrow?" Code *.

                        Teacher replies, "Chris, don't shout out questions." Code –.

Fig. 4.1. Interaction codes (from Stallings, Needels, and Sparks 1987)

To record the interactions for a class with the interaction codes by Stallings, Needels, and Sparks (1987), the observer uses the seating chart that lists the name, gender, and race of each student. Figure 4.2 shows how interaction codes are marked on the seating chart. For example, the black females in this class are Kay, Olga, Hazel, and Robin; the white males are Francis, Bob, David, Ken, Evan, and Tom.

Note that Hazel had no individual interactions with the teacher. Jill was asked a knowledge-level question; she was praised or acknowledged. Later Jill was asked a higher cognitive question and was praised or acknowledged. Then Jill asked a question of the teacher. In response, the teacher asked Jill a knowledge-level question. Jill's answer was wrong or incomplete, and the teacher guided her to another answer. Note: Teachers often address questions to the entire class. To code these whole-group interactions, the observer uses the box designated "Class" at the top of the seating chart. Also, note that the teacher interacted most with students who sat close to the desk and that the majority of the reprimands for behavior were to students sitting in the back row.

Teacher's desk		Chalkboard		
		Class ?		
? + ? G + * ? + * Sarah FW	? + H + * ? G Jill FW	S ? C ? + * Kay FB	S * + Tim MB	* + Olga FB
? + ? + * + John MB	? C ? C * + George MB	S H + Francis MW	H Bob MW	 Shirley FW
? C ? + David MW	 Hazel FB	* + Robin FB	? Elaine FW	– – Ken MW
? C – Evan MW	* – ? + Tony FW	– ? C Bill MB	? C Ellen FW	– – Tom MW

Interaction Seating Chart to analyze a 50-minute period.

FB = Black Female   FW = White Female

MB = Black Male   MW = White Male

Fig. 4.2. Interaction seating chart (from Stallings, Needels, and Sparks 1987)

Figure 4.3 presents some questions for summarizing the interactions. It is important to summarize the number and type of individual interactions the teacher had with black females, white females, black males, and white males, which allows the teacher to analyze patterns by both gender and race.

AFFECTIVE ISSUES

Although the primary goal of mathematics teachers is for students to learn and be able to use mathematics, another important outcome is for students to develop positive attitudes toward mathematics and themselves as learners of mathematics. In addition, students' attitudes have been identified as factors that help explain gender differences in mathematics achievement and course taking (e.g., Leder 1992). Research has identified a number of affective variables as important in understanding the mathematics achievement patterns of females and males. Among these variables are confidence in one's ability to learn mathematics, perceived usefulness of mathematics, and sex-role congruency of mathematics. It is important to recognize that these "variables overlap and interact, both conceptually and in the ways they are typically operationalized and measured" (Leder 1992, p. 614). We distinguish among them as a descriptive convenience.

Confidence in one's ability to learn mathematics is one of the most important attitudinal variables (Fennema and Sherman 1977, 1978; Meyer and Koehler 1990; Reyes 1984; Sherman and Fennema 1977). It is positively correlated to mathematics achievement at middle school and high school levels and is a predictor of students' decisions to enroll in elective mathematics courses in high

How many students were in the class? _____

How many students were spoken to as individuals? _____

Where was the student most spoken to sitting? _____

Where were the students not spoken to sitting? _____

Was there any pattern?

Number of ...	Class	Individuals				Total
		FB	FW	MB	MW	
Direct questions asked?						
Checks for understanding asked?						
Higher cognitive questions asked?						
Guides?						
Corrections?						
Praises?						
Reprimands?						

Fig. 4.3. Summary of interaction observation (from Stallings, Needels, and Sparks 1987)

school. In the classic work of Fennema and Sherman (Fennema and Sherman 1977, 1978; Sherman and Fennema 1977), gender differences in mathematics achievement in favor of males were accompanied by gender differences in confidence, with males more confident than females. Even when there were no differences in mathematics achievement, there were gender differences in confidence, with males reporting greater confidence in their ability to learn mathematics than females reported. In both middle school and high school, females were less confident than males. Further, confidence had the strongest positive correlation with mathematics achievement of all the affective variables in the Fennema-Sherman studies ($r = .40$). Researchers and reviewers since Fennema and Sherman have found similar patterns (e.g., Eccles 1983; Meyer and Koehler 1990; Reyes 1984).

Student perceptions of the usefulness of mathematics is an affective variable that has been shown to be related both to students' mathematics achievement and to their decisions to enroll in elective mathematics courses in high school (Meyer and Koehler 1990). After they have met the high school mathematics requirement, students are unlikely to continue to enroll in mathematics courses unless they see that the mathematics they learn will be of use to them. Fennema and Sherman (1977, 1978) found that gender differences in mathematics achievement in favor of males were accompanied by gender differences in perceived usefulness, also in favor of males. They also found that perceived usefulness was the strongest affective predictor of plans to enroll in elective high school mathematics courses.

Sex-role congruency has to do with the degree to which a student views mathematics as an appropriate activity for his or her gender. Although sex-role congruency is theoretically important in understanding gender and mathematics education, research about its relationship to mathematics achievement and participation has produced inconsistent results (Meyer and Koehler 1990). The degree to which female and male students, teachers, parents, and other members of the community stereotype mathematics as a male domain can have an important impact on girls' willingness to pursue mathematics as an area of interest, particularly during middle school and high school. If girls are concerned that excelling in mathematics is not a sex-role appropriate activity, they may choose not to focus their efforts on learning mathematics. Equally important, if girls perceive that boys view mathematics as a more appropriate activity for males than for females, girls may reduce the value they place on studying and achieving in mathematics. Thus, sex-role congruency for both girls and boys is important to an understanding of students' attitudes about themselves as learners of mathematics.

Rethinking the General Findings

Confidence, usefulness, and sex-role congruency have been identified as important affective characteristics of both male and female mathematics students. Stanic and Hart (1995) studied attitudes toward mathematics and concluded that it was difficult to isolate the effects of a particular attitude because of the way in which attitudes seemed to interact with each other. Their interviews with Cathy, a seventh-grade student, led them to conclude that Cathy's relatively high level of confidence interacted with her lack of enjoyment of mathematics and her statements that mathematics had limited usefulness to her. They found that paper-and-pencil attitude instruments gave only a small part of the picture of Cathy's attitudes. Their interview with her and their observations of her during her mathematics class were particularly important in understanding how her attitudes were related to her achievement in mathematics.

Action Research Ideas

Affective variables are often quite subtle, but they can be very influential in how students perform in mathematics class and what they actually remember. Exploring affective variables related to mathematics offers insights into individuals, but it also can influence daily instruction for the group. For example, if students perceive mathematics as a predominantly male domain, classroom interaction, individual motivation and perseverance, and even achievement are all influenced by the perception. Even when an individual girl does not see mathematics as a male domain, she may be aware and concerned that many boys do see it as a male domain.

Action research could include asking students to portray a mathematician and then explain the important features of their portrait. Young students can draw the mathematician; older students may want to submit a written portrait. Be cautious in giving instructions. Do not refer to the mathematician as he or she. The research not only should provide information about the perceived gender of mathematicians but, more important, should describe the characteristics of a mathematician. What gender patterns exist? How do the characteristics fit gender stereotypes?

Action research can focus on learning more about the range of mathematical confidence of students. As a homework assignment, assign the task shown in figure 4.4 and then ask students to discuss their results with a partner of their choice. Collect the student tasks and compare their confidence domains.

We know that many students agree that mathematics is useful, but they struggle to offer nontrivial examples of useful mathematics. Have students spend ten minutes listing all the ways in which mathematics can be used. Instruct students to be specific. Examine the responses to determine the diversity and significance of their ideas. When teaching various topics throughout the year, emphasize how the mathematics can be used. Post a list of the uses of mathematics and have students add to it as they learn about new topics.

The line below represents your range of confidence. A 0 indicates low confidence and a 10 indicates high confidence. Place a star on the line to show how confident you are about doing mathematics in general.

Below the line, explain what it is about mathematics that caused you to place your star where you did.

| 0 | 1 | 2 | 3 | 4 | 5 | 6 | 7 | 8 | 9 | 10 |

Explanation:

I am confident about _____

I am not confident about _____

Fig. 4.4. Research activity on mathematical confidence

Journal entries about mathematics and about doing mathematics are a great source of data for action research. On separate days, invite a male and a female speaker to class to talk about how they use mathematics in their jobs. After each speaker, give students a journal writing prompt, such as "Picture yourself working in the same job as our speaker. What mathematics do you need to know? Would you like to do this job at some time in your life?" Do boys identify more with the male speaker and girls with the female speaker? Can both boys and girls recognize mathematics in each job? It would be particularly interesting to have speakers with nontraditional gender roles, such as a male nurse and a female construction worker.

CLOSING THOUGHTS

We have suggested some areas in which the action research findings may be particularly helpful to teachers making instructional and curricular decisions. Equally important, the actual process of collecting data, regardless of findings, strengthens instruction. The patterns that teachers identify through action research can help them better understand their students and the social context in which they teach mathematics. The process of conducting research may make both teachers and students more aware of gender issues. Discussing the research with students can help them become more aware of their own identity and of how they think about learners of a different gender.

Although research can create exciting environments in individual classrooms, it can also make an impact on a school system and a community. Sharing action research projects with colleagues, administrators, and parents can promote gender equity. Creating a working group in the community to study gender issues in mathematics has the potential of helping everyone understand the importance of equity in mathematics education. In *Failing at Fairness*, the Sadkers (1994) discussed the educational traumas that girls still endure and how subtle and yet pervasive the inequities are in the educational system. They also discussed the potential of the educational system to make a difference:

> When all these citizens from our American village join forces, they can transform our educational institutions into the most powerful levers for equity, places where girls are valued as much as boys, daughters are cherished as fully as sons, and tomorrow's women are prepared to be full partners in all activities of the next century and beyond. (P. 280)

As mathematics teachers and action researchers, we have the potential not only to understand equity issues in mathematics education but also to make a difference in the mathematics education of our students—all our students.

REFERENCES

Becker, Joanne Rossi. "Differential Treatment of Females and Males in Mathematics Classes." *Journal for Research in Mathematics Education* 12 (January 1984): 40–53.

Brandon, Paul R., Barbara J. Newton, and Ormond W. Hammond. "Children's Mathematics Achievement in Hawaii: Sex Differences Favoring Girls." *American Educational Research Journal* 24 (Fall 1987): 437–61.

Brophy, Jere E. "Interactions of Male and Female Students with Male and Female Teachers." In *Gender Influences in Classroom Interaction*, edited by Louise Cherry Wilkinson and Cora B. Marrett, pp. 115–42. Orlando, Fla.: Academic Press, 1985.

College Board. *College-Bound Seniors: A Profile of SAT Program Test Takers.* New York and Princeton, N.J.: College Entrance Examination Board and Educational Testing Service, 1996.

——. *Taking the SAT I: Reasoning Test*. New York and Princeton, N.J.: College Entrance Examination Board and Educational Testing Service, 1997a.

——. "Common Sense about SAT Score Differences and Test Validity." *Research Notes* (June 1997b): 1–12.

Damarin, Suzanne K. "Gender and Mathematics from a Feminist Standpoint." In *New Directions for Equity in Mathematics Education*, edited by Walter G. Secada, Elizabeth Fennema, and Lisa Byrd Adajian, pp. 242–57. New York: Cambridge University Press, 1995.

Dossey, John, Ina Mullis, Mary M. Lindquist, and Donald Chambers. *The Mathematics Report Card*. Princeton, N.J.: Educational Testing Service, 1988.

Eccles, Jacquelynne S. "Expectancies, Values, and Academic Behaviors." In *Achievement and Achievement Motives*, edited by Janet T. Spence, pp. 75–146. San Francisco: Freeman, 1983.

Fennema, Elizabeth. "Mathematics, Gender, and Research." In *Gender and Mathematics Education*, edited by Barbro Grevholm and Gila Hanna, pp. 21–38. Lund, Sweden: Lund University Press, 1995.

Fennema, Elizabeth, and Penelope L. Peterson. "Teacher-Student Interactions and Sex-Related Differences in Learning Mathematics." *Teaching and Teacher Education* 20, no. 1 (1986): 19–42.

——. "Effective Teaching for Girls and Boys: The Same or Different?" In *Talks to Teachers*, edited by David C. Berliner and Barak V. Rosenshine, pp. 111–25. New York: Random House, 1987.

Fennema, Elizabeth, and Julia Sherman. "Sex-Related Differences in Mathematics Achievement, Spatial Visualization, and Affective Factors." *American Educational Research Journal* 14 (Winter 1977): 51–71.

——. "Sex-Related Differences in Mathematics Achievement and Related Factors: A Further Study." *Journal for Research in Mathematics Education* 9 (May 1978): 189–203.

Hart, Laurie E. "Classroom Processes, Sex of Student, and Confidence in Learning Mathematics." *Journal for Research in Mathematics Education* 20 (May 1989): 242–60.

Hopkins, David. *A Teacher's Guide to Classroom Research*. Philadelphia: Open University Press, 1985.

Jones, Lyle V. "The Influence on Mathematics Test Scores, by Ethnicity and Sex, of Prior Achievement and High School Mathematics Courses." *Journal for Research in Mathematics Education* 18 (May 1987): 180–86.

Kimball, Meredith M. "A New Perspective on Women's Math Achievement." *Psychological Bulletin* 105 (March 1989): 198–214.

Lankford, Nina K. "Teachers as Researchers: What Does It Really Mean?" In *Research Ideas for the Classroom: High School Mathematics*, edited by Patricia S. Wilson, pp. 279–89. New York: Macmillan, 1993.

Leder, Gilah C. "Gender Differences in Mathematics: An Overview." In *Mathematics and Gender*, edited by Elizabeth Fennema and Gilah C. Leder, pp. 10–26. New York: Teachers College Press, 1990.

——. "Mathematics and Gender: Changing Perspectives." In *Handbook of Research on Teaching and Learning Mathematics*, edited by Douglas A. Grouws, pp. 597–622. New York: Macmillan, 1992.

Lytle, Susan L., and Marilyn Cochran-Smith. "Teacher Research as a Way of Knowing." *Harvard Educational Review* 62 (Winter 1992): 447–74.

Meyer, Margaret R. "Gender Differences in Mathematics." In *Results from the Fourth Mathematics Assessment of the National Assessment of Educational Progress*, edited by Mary M. Lindquist, pp. 119–59. Reston, Va.: National Council of Teachers of Mathematics, 1989.

Meyer, Margaret R., and Mary Schatz Koehler. "Internal Influences on Gender Differences in Mathematics." In *Mathematics and Gender*, edited by Elizabeth Fennema and Gilah C. Leder, pp. 60–95. New York: Teachers College Press, 1990.

Mullis, Ina, Eugene Owen, and G. Phillips. *Accelerating Academic Achievement*. Princeton, N.J.: National Assessment of Educational Progress, Educational Testing Service, 1990.

National Center for Education Statistics (NCES). *Digest of Educational Statistics*. Washington, D.C.: U.S. Department of Education, 1995.

Reese, Clyde M., Karen E. Miller, John Mazzeo, and John A. Dossey. *NAEP 1996 Mathematics Report Card for the Nation and the States*. Washington, D.C.: National Center for Education Statistics, 1997.

Reyes, Laurie Hart. "Affective Variables and Mathematics Education." *Elementary School Journal* 84 (May 1984): 558–81.

Reyes, Laurie Hart, and George M. A. Stanic. "Race, Sex, Socioeconomic Status and Mathematics." *Journal for Research in Mathematics Education* 19 (January 1988): 26–43.

Sadker, Myra, and David Sadker. *Failing at Fairness: How America's Schools Cheat Girls*. New York: Charles Scribner's Sons, 1994.

Schildkamp-Kundiger, Erica, ed. *International Review on Gender and Mathematics*. Columbus, Ohio: ERIC Clearinghouse for Science, Mathematics, and Environmental Education, 1982.

Sherman, Julia S., and Elizabeth Fennema. "The Study of Mathematics among High School Girls and Boys: Related Factors." *American Educational Research Journal* 14 (Spring 1977): 159–68.

Silver, Edward A., Marilyn E. Strutchens, and Judith S. Zawojewski. "NAEP Findings Regarding Race/Ethnicity and Gender: Affective Issues, Mathematics Performance, and Instructional Context." In *Results from the Sixth Mathematics Assessment of the National Assessment of Educational Progress*, edited by Patricia Ann Kenney and Edward A. Silver, pp. 33–59. Reston, Va.: National Council of Teachers of Mathematics, 1997.

Stallings, Jane A., Margaret Needels, and Georgea Mohlman Sparks. "Observation for the Improvement of Classroom Learning." In *Talks to Teachers*, edited by David C. Berliner and Barak V. Rosenshine, pp. 129–58. New York: Random House, 1987.

Stanic, George M. A., and Laurie E. Hart. "Attitudes, Persistence, and Mathematics Achievement: Qualifying Race and Sex Differences." In *New Directions for Equity in Mathematics Education*, edited by Walter G. Secada, Elizabeth Fennema, and Lisa Byrd Adajian, pp. 258–76. New York: Cambridge University Press, 1995.

Stanic, George M. A., and Laurie Hart Reyes. "Excellence and Equity in Mathematics Classrooms." *For the Learning of Mathematics* 7 (June 1987): 27–31.

Assessing Achievement in Mathematics

Eliminating the Gender Bias

5

Charles E. Mitchell

Ann Calahan

Does bias against females continue to exist in the teaching of mathematics in the classrooms of the United States? Does the deleterious impact of gender bias help explain the differences in the performances of males and females on standardized tests of performance in mathematics, such as the PSAT and SAT exams? Representatives of the Educational Testing Service (ETS) (Kelly 1994) acknowledge that males outscore females in mathematics by an average of 45 points on the SAT exam (Manning 1994) but deny that the differences are due to the effects of gender bias. These exams are often a factor in determining the recipients of academic scholarships. Their use in this respect often results in a final pool of applicants that contains a disproportionate number of males. Yet summaries of standardized test scores conflict with other data suggesting that females have higher grades in mathematics courses both in college and in high school (Kelly 1994; Sadker and Sadker 1994). Add to this perspective the numerous investigations of gender bias over the past two decades. Their reports reveal clear evidence that gender bias continues to be a serious problem both in how mathematics is taught and how mathematical performance is evaluated.

GENDER ISSUES IN ASSESSMENT PRACTICES

A gap between the way that mathematics is taught and the way that performance is later evaluated may seriously obstruct efforts to provide an equitable educational experience for all students. The practice of constructing traditional examinations in which the time allotted to complete the exam is a major factor may create unfair and unreasonable obtacles for female students. In an investigation by Miller, Mitchell, and Van Ausdall (1994), 139 students enrolled in advanced mathematics classes (second-year algebra, trigonometry, precalculus) from four high schools were administered thirty-minute SAT-like practice exams on two separate occasions. The tests were commercially available and designed to give students an idea of what they would face when they took the SAT exam and how they might score. In one session, the students were limited to the suggested time of thirty minutes. In another session, the students were told that they had as much time as they needed. Whether the students were administered the timed or the untimed test first was randomly assigned. The overall performance of males and females was evaluated with traditional parametric t tests. An initial analysis of the scores from both sessions suggested that males had significantly outscored females on both administrations of the examinations. However, when time was removed as a factor and t tests were applied to these data, the scores of females significantly improved, whereas the scores of males did not. For this study, moderating the time factor allowed more equitable outcomes in mathematics.

To educators familiar with research studies on gender bias, the possibility that the length of time that students are given to take a test can produce biased results favoring males will not be surprising. Investigations suggest strongly that females are conditioned to be neater in their work than males (Macoby and Jacklin 1974). It is reasonable to assume that trying to be neat consumes extra time. Females are also reported to be more inclined than males to choose lengthier and more detailed solution methods (Macoby and Jacklin 1974), whereas males are more inclined to size up an item quickly and to estimate a solution (Hudson 1986). If time is a major factor on an examination, judiciously using educated guessing strategies may be superior to employing complicated algorithms that lead directly to a correct solution, particularly if the student must think about choosing and implementing the algorithm.

Another problem-solving strategy more commonly employed by females than by males may increase the overall negative impact of time on the performance of females. Females were reported to be more inclined than males to draw pictures to help them find a solution to an item (Fennema and Tartre 1985). This practice again consumes more time and could exacerbate a student's dilemma if time already is a factor.

Mitchell (1996) conducted a series of action studies with two fourth-grade and two sixth-grade teachers. The teachers were asked to construct twenty-item examinations for their students by using items drawn directly from state assessment inventories and the four-option, multiple-choice format employed on the state examination. The administration of the tests to the students differed in only two ways: the teacher-constructed examinations were shorter than the state inventories, and the students were asked to submit any work they used to compute a solution. The teachers then used the state's penalty for incorrect responses as they scored the completed exams. In all four classes, the males outscored the females. The teachers were then asked to rescore the examinations. On the second scoring, the teachers were asked to give a student one-half credit for an item if evidence from the worksheets suggested that he or she had employed a viable solution method. The teachers were further asked to eliminate credit for an item if the item was answered correctly but the student's work did not support the solution. In other words, if a student employed an invalid solution strategy but then correctly guessed the correct response, no credit was given for the item. The examinations were reevaluated. In all four classes, the scoring summaries favored the females over the males. A summary of the time-associated practices that may have influenced the students' scores follows:

- When recopying an item to a worksheet, girls tend to more carefully include all mathematical symbols, such as plus or minus signs. (Fourth grade)
- Girls are more inclined to write out solutions in concrete terms, such as "36 cookies." (Fourth grade)
- Boys are more inclined to miscopy an item when transferring it to a worksheet. (Fourth grade)
- Girls are more inclined to show evidence that they considered a variety of solution methods when solving an item. (Sixth grade)
- Girls are more inclined to show evidence that they checked the solution to an item. (Sixth grade)
- When multiple-choice exams based on state assessment items were scored traditionally, boys tended to score higher than girls. When the same exams were rescored to eliminate credit for guessing and to add partial credit where it was earned, girls tended to score higher than boys. (Fourth and sixth grades)
- In open-classroom situations, girls were more inclined than boys to read an item twice. (Sixth grade)

- In open-classroom situations, girls were more inclined to show evidence that they had estimated the final solution to an item before solving it. (Sixth grade)

It is interesting that these procedures and problem-solving strategies are all desirable educational objectives, yet they all consume additional time. If time is a major factor in an exam, as it certainly is on such standardized examinations as the SAT, and females are more inclined than males to practice the foregoing desirable problem-solving strategies, then the question of bias and unfairness becomes clear.

SUGGESTIONS FOR ELIMINATING BIAS FROM ASSESSMENT PRACTICES

When evaluating students, teachers need to address both the construction of the items to be administered and the establishment of the environment in which students will be assessed. If time is a potentially biasing element of the process, as previous research suggests, teachers should experiment with fewer items on each assessment, thus allowing more time for a student to solve each item. Perhaps instead of three or four items that address the same or similar instructional goals, one or two items would be sufficient. Instead of ensuring that all types of items are included for each assessment activity, a sample of the types may be sufficient.

Students who have been conditioned to work through assessment items hastily may need a period of adjustment with reminders and encouragement from their teacher to review the items and check their work. By giving students more time to think, the teacher may be better able to address the goal of turning the evaluation process into an opportunity for students to learn or understand the material better (National Council of Teachers of Mathematics [NCTM] 1995).

If a student submits a completed exam early, the teacher might review the worksheet and the final responses. If the teacher determines that problems exist with some items, he or she might ask some leading questions before returning the assessment activity: "How would drawing a picture help you solve this item?" "How might you model this item if you used manipulatives?" "What strategies for solving this item did we discuss in class?" "Give me a step-by-step account of your solution method." Such questions might help the student better organize the solution method, without giving away too much information. After all, the purpose of an evaluation is to assess a student's knowledge or understanding of some mathematical content. If a student cannot remember how to get started, the teacher would have difficulty determining what the student knows and would have no opportunity to award deserved partial credit for what the student has accomplished.

Teachers must also become more involved in alternative forms of evaluation, such as written reports, oral presentations, and small-group exercises. Involvement in a rich variety of assessment activities gives each student a better opportunity to exhibit unique qualities, knowledge, and experiences (NCTM 1995). To be fair, assessment activities can no longer ignore differences in gender or in cultural and social backgrounds (NCTM 1995). If female students are more inclined to choose more-detailed solution methods, then giving more opportunities to verbalize both orally and in writing may provide instructors with a better idea of a student's depth of understanding.

Other changes in the evaluation process may be necessary to achieve equity. Research suggests that both females and minority students perform better when the relevance of the material is made clear (Brown 1986). Assessment instruments should be constructed to reflect the instructional activities in which the content was introduced and developed (NCTM 1995). Clearly, the real-life relevance of the content should be at the heart of instruction (NCTM 1989). Thus, a greater percentage of the assessment items should be word problems similar to those used to introduce and develop content. Indeed, all students should benefit from learning how mathematics can be used in various occupations or in every-day life situations.

Educators have made a great deal of progress in controlling blatant forms of gender bias. However, much work is still needed to understand the complex nature of gender bias and then to take appropriate steps to eliminate more subtle forms of bias. To accomplish these goals, future research efforts must establish whether significant differences exist in the way that males and females approach examinations and how these differences might negatively affect either gender. Establishing an equitable educational opportunity for all students will require the combined efforts of teachers and parents. An increasing awareness of what constitutes gender bias, and the damage currently inflicted on large groups of students, should be all the incentive necessary. The reward is a more equitable opportunity for an education for all students.

REFERENCES

Brown, Thomas J. *Teaching Minorities More Effectively: A Model for Educators.* New York: University Press of America, 1986.

Fennema, Elizabeth, and Lindsay A. Tartre. "The Use of Spatial Visualization in Mathematics by Girls and Boys." *Journal for Research in Mathematics Education* 16 (1985): 184–206.

Hudson, Lisa. "Item-Level Analysis of Sex Differences in Mathematics Achievement Test Performances." (Ph.D. diss., Cornell University, 1986.) *Dissertation Abstracts International* 47 (1986): DA8607283.

Kelly, Dennis. "Suit Calls National Merit Test Biased against Girls." *USA Today,* 16 February 1994, p. A1.

Macoby, Eleanor E., and Carol Jacklin. *The Psychology of Sex Differences.* Stanford, Calif.: Stanford University Press, 1974.

Manning, Anita. "How Bias in Coed Classrooms Holds Girls Back." *USA Today,* 2 February 1994, p. D5.

Miller, L. Diane, Charles Mitchell, and Marilyn Van Ausdall. "Evaluating Achievement in Mathematics: Exploring the Gender Biases of Timed Testing." *Education* 114 (Spring 1994): 436–39.

Mitchell, Charles. "Gender Equity and the Multiple Choice Exam." *Texas Mathematics Teacher* 43, no. 2 (1996): 3–4.

———. *Assessment Standards for School Mathematics.* Reston, Va.: National Council of Teachers of Mathematics, 1995.

National Council of Teachers of Mathematics. p*Curriculum and Evaluation Standards for School Mathematics.* Reston, Va.: National Council of Teachers of Mathematics, 1989.

Sadker, Myra, and David Sadker. *Failing at Success: How Our Schools Cheat Girls.* New York: Simon & Shuster, 1994.

Crucial Points in Mathematics Decision Making

Advice for Young Women

6

At crucial mathematics decision points, parents and their children make decisions that affect the children's mathematics future—and their economic future as well. Believe it or not, these decisions start as early as the fifth grade. Here is some information for parents and advice for their children.

Patricia B. Campbell

Kathryn B. Campbell-Kibler

FIFTH AND SIXTH GRADE: DO I ACCELERATE?

Parent Information

Decisions made in fifth and sixth grade influence whether your child can accelerate in mathematics—that is, take first-year algebra in eighth grade—and have the potential to go on to take calculus. Saying yes to accelerating in mathematics does not mean that your child has to take calculus or even geometry, but it does mean that she or he will continue to have more choices.

Girls may confront special issues at this point. Some people, even some teachers, still tend not to see girls as being good in mathematics. It is often very difficult for elementary school teachers to predict how well a child will do in algebra and geometry on the basis of how well she or he has done in arithmetic. Stereotyped expectations about who can, and should, do well in mathematics can influence decisions.

Advice for Fifth and Sixth Graders

You do not have to be a mathematics whiz or a science nerd to accelerate in mathematics. Most of the mathematics studied in seventh and eighth grade is arithmetic—some variation of addition, subtraction, multiplication, and division. With algebra, mathematics becomes a lot more interesting and fun. It can even be creative; sometimes you can solve problems by using your own methods and ideas. So unless there is a good reason not to accelerate, why not go for it? Take algebra in eighth grade!

If you have not been chosen to accelerate but would like to, see whether your parents agree. In many states, parents can override the teacher's and the school's decision about placement in mathematics.

EIGHTH GRADE: DO I TAKE ALGEBRA NEXT YEAR?

Parent Information

Children who do not accelerate in seventh grade need to decide in eighth grade whether they will take algebra or general mathematics in ninth grade. There is, or should be, no real choice: strongly encourage them to take algebra.

Advice for Eighth Graders

If you did not take algebra in eighth grade, you will need to decide whether to take algebra or a general mathematics course in ninth grade. There is, or should be, no real choice: Take algebra!

Without algebra you cannot continue in mathematics; cannot get into most colleges, even junior or community colleges; and are severely limited in the jobs you can get. Even such supposedly nonmathematics jobs as nursing, computer repair, and sales require algebra. Increasing numbers of schools and cities think that algebra is so important that they require every student, other than special education students, to take it. If your school does not require algebra, take it anyway. You are probably already doing some algebra without knowing it.

Remember, regardless of what you want to do in life, take algebra. It will not hurt, and it most likely will help a lot.

TENTH AND ELEVENTH GRADE: DO I GO ON?

Parent Information

The next mathematics decision-making time occurs when a student has taken three years of mathematics—the traditional amount necessary to get into a good college. Students then have to decide whether they want to go on to fourth-year mathematics and beyond. Accelerated students make this decision in tenth grade; others make it in eleventh. Your child should seriously consider going on in mathematics.

Advice for High School Students

Are you thinking about whether to take a fourth year of mathematics? Well, seriously consider these reasons for going on:

- Three years of high school mathematics is the minimum for most good colleges, and some require four. Regardless of the minimum requirement, if you have more years of mathematics, you have a better chance of being admitted regardless of your major.
- Colleges—and, increasingly, high schools—no longer penalize students who take harder courses. An OK grade in an advanced mathematics course counts more than an A in an undemanding course.
- The more mathematics you take, the broader your future career options will be. Having too much mathematics will not stop you from being anything that you want to be; having too little will.
- Particularly for girls, taking more mathematics means making more money. In 1991, a United States government study found that women and men who took eight credits of college mathematics (usually calculus) made more money than college graduates who did not. Although men usu-

ally make more money than women in the same jobs, this occurs less often in mathematics-related jobs. In fact, in some mathematics-related jobs like engineering, on the average, women under thirty make more money than men of the same age!

Besides, if you do not take mathematics in your senior year, you may forget some of it. Then when you take a placement exam in mathematics at your college, you may do so poorly that you have to repeat some of the high school mathematics you once knew.

Parent Information

At the final mathematics decision point in high school, accelerated students must decide whether they are going to take calculus in high school and if so, whether they will take the Advanced Placement (AP) exam. If a student takes the AP exam and scores well, most colleges will give college credit for the course.

Advice for Seniors

Are you thinking about taking calculus in your senior year? If so, you should take Advanced Placement calculus and the AP exam. Whether you decide to take calculus in high school is really your call. If you are planning to go into engineering or another mathematics- or science-related career or if you are planning to major in mathematics or physics or chemistry in college, then of course you should take it. You also might want to take it so that you will then have completed your college mathematics requirement. When you get to college, you will then have time for more electives. Also, calculus does make you more competitive in college applications. If you are not sure what to do, talk to the calculus teacher and see what the course is going to be like. Who knows? It may be fun.

Advice For Everyone

Young women tend to be underrepresented in advanced mathematics courses, particularly calculus—and not because they do not have the skills to do it. Sometimes all that they need is extra encouragement. That encouragement can come from teachers, parents, or even one another.

More mathematics means more choices.
Mathematics can make a difference in your life.

ADVANCED PLACEMENT COURSES: WHY ME?

The Mathematics and Mathematical Thinking of Four Women Seamstresses

Sabrina J. C. Hancock

7

As a young girl, I visited Grandma every summer at her home in the Mississippi delta. Nothing fascinated me more than watching her take beautiful pieces of fabric and shape them into the hottest fashions of the season.

The art of sewing was ritually passed down by the women in my family, generation to generation. One summer, Grandma decided that I was mature enough to learn to sew. I began to learn how to use the tools, terminology, rules of thumb, and resources available to seamstresses.

Not until I became a mathematics teacher did I reflect on what I had learned as a youth. I realized that I used much geometry and measurement in sewing. This awakening transformed my ideas about mathematics and its creators. Until then I had viewed mathematics as a set of universal truths rather than as a cultural product. I thought that the creators of mathematics were mostly men, such as Newton and Fermat, who were referred to in my classroom mathematics books, rather than everyday people. I realized that many of us experience, use, and create mathematics every day; however, we do not always recognize the mathematics in what we are doing.

This study focuses on the mathematics used daily by four women in the context of sewing. It describes the mathematics that I recognize and documents the mathematics that the seamstresses recognize in their skills, thinking, and strategies.

RATIONALE AND THEORETICAL FRAMEWORK

If asked to name five female mathematicians, many of us could not do it. Many of us could, however, name five male mathematicians. This should not be much of a surprise because many of our school mathematics books are dominated by contributions made by European men.

Although more women are participating in, and contributing to, the field of academic mathematics today, many capable women still choose not to pursue careers in mathematics. One explanation is that mathematics is viewed as a male domain; therefore, pursuing a career in mathematics is not consistent with a woman's gender identity. To create equity in mathematics education, women must perceive mathematics as being valuable to their lives as well as being appropriate for their gender identity. Mathematics must be viewed as a female as well as a male domain.

For mathematics to be viewed as being appropriate for women, the mathematics curriculum must included their contributions, which, unfortunately, have often been ignored (Fasheh 1982; Harris 1987a).

> Math was necessary for [my mother] in a much more profound and real sense than it was for me. My illiterate mother routinely took rectangles of fabric and, with few measurements and no patterns, cut them and turned them into beautiful, perfectly fitted clothing for people. In 1976 it struck me that the mathematics she was using was beyond my comprehension; moreover while mathematics for me was a subject matter I studied and taught, for her it was basic to the operation of her understanding . . . . Seeing my mother's math in context helped make me see my math in context; the context of power. What kept her from being fully a praxis and limited her empowerment was a social context which discredited her as a woman and uneducated and paid her extremely poorly for her work (Fasheh 1991, p. 57).

Because his mother's mathematics did not fit into the Western world of mathematics, it "was continuously discredited by the world around her, by the culture of silence and cultural hegemony" (Fasheh 1991, p. 59). The mathematics and intellect that should have empowered Fasheh's mother were never acknowledged.

Similarly, Harris (1987a, 1987b) argued that women's contributions to mathematics have been ignored. She compared a geometrical figure to a motif from a woven Turkish rug. Although the figures appeared identical, the motif was much more difficult to construct because it had to be woven in relation to the other motifs on the symmetrical rug.

> For reasons which need to be closely examined, figure 1 [the geometrical figure] seems to count as mathematics, figure 2 [the motif] does not. Most of the reasons suggested to the writer so far do not stand up to much examination. A summary of them is that figure 2 [a weaving] is simply not taken seriously as mathematics because firstly the weaver has had no schooling and is illiterate and, secondly, she is a girl. It has even been stated outright by more than one mathematics educator that the weaver "is not thinking mathematically", to which the immediate response must be "How do you know? Have you asked her what her thinking was?" (Harris 1987b, p. 27)

A more recent explanation of why women are not pursuing careers in mathematics is that many prefer to learn in a way that is not consistent with the way that mathematics is traditionally taught in school. Many of these arguments are based on the framework developed by Belenky, Clinchy, Goldberger, and Tarule (1986). These researchers found that most women in their study preferred to learn in a connected manner—a way of knowing that builds on the women's conviction "that the most trustworthy knowledge comes from personal experience rather than the pronouncements of authority" (p. 113).

Mathematics educators have argued that traditional mathematics is typically taught in a separate manner—a manner that embodies "logic, deduction, and certainty" (Becker 1994, pp. 16–17). Buerk (1985) wrote that separate mathematics involved solving the problem in a structured algorithmic way while stripping away any context, whereas connected mathematics involved experiencing the problem and connecting it to the personal world. She also wrote that although women may prefer to learn in a connected manner, they are able to master the separate way of reasoning and to achieve in mathematics. Therefore, even the women who achieve in academic mathematics may not be motivated to pursue mathematics-related careers because their goals and ways of reasoning may differ from those espoused in mathematics classrooms.

I believe that some women have failed to pursue mathematics partly because of the narrow definition of legitimate mathematics and mathematical thinking. Many people believe that mathematics is used and created only by an elite group of specialists, such as mathematicians, engineers, and physicists. Everyday people rarely see themselves as users and creators of complex mathematics. Lave (1985) noted that it was not unusual for individuals to apologize for not using "real" mathematics as they relied on their own strategies to solve mathematical problems.

Researchers have shown that different goals, tools, thinking, and resources cause mathematics to develop in different directions (Bishop 1988; D'Ambrosio 1985; Gerdes 1985; Millroy 1992). This view, that mathematics is shaped by the tools and thinking of a culture, contrasts with the narrow view that only one true mathematics exists among all cultures. Researchers who have argued for a broader view of mathematics have studied the mathematics and mathematical thinking of cultural groups. These groups include carpenters (Millroy 1992), carpet layers (Masingila 1992), fishers (Gerdes 1985), weavers (Harris 1987b), dairy truck drivers (Scribner 1984), dieters (de la Rocha 1985), tailors (Lave 1977; Petitto 1979), and grocery shoppers (Murtaugh 1985).

Unfortunately, few studies richly describe the mathematics practiced outside of school, and most of these studies focus on traditional male domains, such as carpentry and carpet laying. It is equally important to document the mathematics practiced by women.

As mathematics teachers, we need to learn more about alternative ways of knowing and applying mathematics so that we can modify our teaching to include the interests, goals, and ways of reasoning of others. We need to draw on legitimate mathematics applications that are not limited to a few contexts and professions.

In summary, this study is important for at least two reasons. First, it documents women as participants in, and creators of, mathematics. Second, it identifies legitimate mathematical applications that are used by women in a traditional female context.

Methodology

I used an ethnographic approach to analyze and describe the seamstresses' mathematics and mathematical thinking. For this study, I spent 37.5 hours with four seamstresses over four months. Participant observation, informal and semistructured interviews, and researcher introspection were the techniques I used to collect data. Interviews were recorded by audiotaping, videotaping, and notetaking. I used scanning, coding, categorizing, defining, and synthesizing to find relationships among the data. I arrived at relationships inductively and then tested them by looking for disconfirming evidence.

My mathematical training gave me a framework for recognizing the mathematics that resembled school mathematics. Because I have sewn for the past ten years, I was familiar with many of the seamstresses' tools, symbols, and metaphors. Therefore, I used the framework of a seamstress to help me recognize the seamstresses' nontraditional mathematics.

THE SEAMSTRESSES

Four professional seamstresses who have sewn for the public for at least four years participated in this study. All the seamstresses were asked how they would like to be addressed in this study, and all but one asked to be called by their first name. Mrs. Bowen asked me to address her as "Mrs. Bowen" probably because many of her clients, colleagues, and friends know her by that name. Two of the seamstresses, Joan and Karen, were recommended by their clients. Christina and Mrs. Bowen were recommended by Joan because they could create intricate garments without a pattern.

Joan, a fifty-one-year-old seamstress, former mathematics teacher, mother, and wife, classifies herself as a pattern sewer because she uses patterns to sew clothes for clients. She learned to sew garments when she was nine years old by observing and asking questions. Joan majored in mathematics and chemistry at a southern college. She taught a variety of middle and high school mathematics courses, which ranged from general mathematics to trigonometry, for eight years. She is not currently teaching.

Christina, a thirty-seven-year-old seamstress, has run her own design business for the past five years. However, she has worked with other designers since she was twenty-three. To create her designs, she makes her own patterns from muslin fabric or changes a purchased pattern. Christina learned to sew when she was about thirteen years old. She went to art school; however, she never graduated from college because she could not pass her mathematics classes although she had taken geometry in high school.

Karen, a thirty-one-year-old wife and mother, has been sewing at home for three years. She has also worked part-time for four years with a fashion designer. Karen learned to sew when she was ten years old by reading pattern directions. She studied graphic design at a southern university until she failed calculus. Karen transferred to a midwestern university, which did not require calculus, and majored in fabric design.

Mrs. Bowen, a seventy-one-year-old mother and wife, learned to sew when she was about nine years old by watching her mother create garments from measurements. She attended public school through the ninth grade; therefore, her exposure to academic mathematics is limited. She sews mostly majorette costumes and pageant dresses.

THE SEAMSTRESSES' MATHEMATICS

The seamstresses used mathematics for estimation, problem solving, measurement, spatial visualization, reasoning, geometry, and cost effectiveness. Because of their different tools, resources, goals, and thinking, their mathematics rarely resembled school mathematics.

Through the seamstresses' language and actions, they exhibited an understanding of the concepts of angles, direction, parallelism, reflection, symmetry, design, proportion, similarity, and estimation. Words such as *bias, dart, nap, pile, straight of the grain, on the fold, enlarge, envision, visual,* and *eyeing it* were used to communicate this mathematical knowledge.

Researchers suggest that different tools, resources, goals, and thinking cause mathematics to develop in different directions (Bishop 1988; D'Ambrosio 1985a; Gerdes 1985; Millroy 1992). Therefore, I examined these components to explain some of the unique characteristics of the seamstresses' mathematics.

Tools and Resources

Some of the tools and resources of the seamstresses were different from those found in schools. Fabric, pattern pieces, and tape measures played significant roles in the development of the seamstresses' mathematics.

The fabric was a two-dimensional plane on which much of the seamstresses' mathematics took place. Unlike traditional mathematics courses, where the plane is often an unmalleable chalkboard, the seamstresses' plane is flimsy and is often folded in half. The fold of the material represents an axis that is also the line of reflection.

The pattern pieces were the seamstresses' geometrical shapes. Just as we are familiar with the axes of symmetry of squares, the seamstresses are familiar with the axes of symmetry of their pattern pieces. The symmetrical pattern pieces were often folded on their axis of symmetry and cut on the fold of the fabric. Pieces that were not symmetrical, but were reflections of each other, such as right and left sleeves, were not cut on the fold; however, the fold often served as the line of reflection.

When the seamstresses did not use a pattern, they used a tape measure to locate points on the material. As seen in figure 7.1, Mrs. Bowen recorded many different measurements of her client. The numbers are the measurements, in inches, of the client's physique. For example, 21-w, which is located on the front side of the costume, refers to the total number of inches around the client's waist. The corresponding measurement on the back, 9, refers only to the number of inches around the back side of the client's waist. Two other measurements listed on the front, 25 1/3 and 23 1/3, refer to the total distances around the client's chest and rib cage, respectively. The rest of the measurements in figure 7.1 refer only to the front half or the back half of the costume.

Most of the measurements that Mrs. Bowen took with her tape measure were parallel or orthogonal; however, there were a few that curved along the body. Mrs. Bowen chalked the measurements onto the fabric by locating the points on her paper and translating these points onto the fabric. For example, when she was cutting the front of a majorette costume on the fold of the fabric, she found the measurement on her tape measure, folded the tape measure in half, and then measured this distance from the fold. As seen in figure 7.2, if the measurement above the girl's bust was 8 1/2 inches, she found 8 1/2 inches on her tape measure, folded the tape measure in half, measured this distance from the fold, and chalked this point. It is interesting to note that Mrs. Bowen did not use calculations to divide 8 1/2 in half; rather, she physically manipulated one of her tools to locate the halfway point.

By transferring the measurements on the paper to the fabric, Mrs. Bowen formed a type of coordinate system on the plane of the fabric. She chose a point on the client to begin measuring. This point was analogous to the origin on a

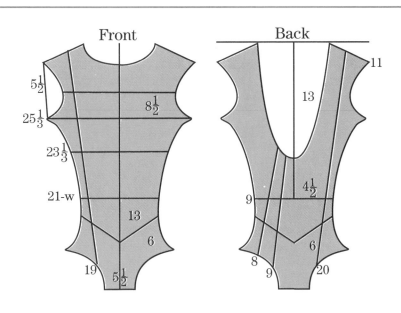

Fig. 7.1. Mrs. Bowen's measurement sheet

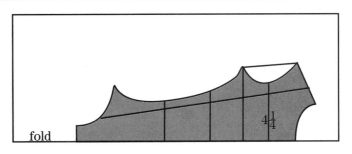

Fig. 7.2. Chalking the points

Cartesian plane. Just as all points can be found with horizontal and vertical movements from the origin in a Cartesian plane, all the client's measurements could be found with respect to the center-front neckline. Mrs. Bowen used her tape measure to measure down and across from the neckline to locate points. She also measured down and across from these new chalk points to locate other points.

Christina used the tape measure as a compass when making a circular skirt. As with a compass, one end of Christina's tape measure stayed fixed on a point as she swung the other end with her arm to create a circular arc.

Although these are not all the tools and resources that the seamstresses used, these tools and resources clearly contributed to the development of their mathematics. The seamstresses were actively engaged with their tools and resources. They folded their plane, reflected their pattern pieces, created a coordinate system, and became part of a compass. Their tools and resources are analogous to mathematical symbols and objects. For example, in school mathematics, perpendicular lines are the symbols used to represent the x-axis and the y-axis of a Cartesian plane. The seamstresses use the fold of the material to represent one axis symbolically while they create the other axis with their measurements. Also, in school mathematics, lines of symmetry of geometrical objects, such as squares and rectangles, are found. The seamstresses found the lines of symmetry of their geometrical objects—the pattern pieces. Therefore, a characteristic of the seamstresses' mathematics is an active engagement with their mathematical symbols and objects.

Goals

Other aspects that contribute to the uniqueness of the seamstresses' mathematics are their goals. The seamstresses share two goals: making a flat piece of fabric fit a three-dimensional person and minimizing time and materials to maximize profit.

From the first goal, spatial visualization and surface geometry developed. When the seamstresses made their own patterns, they had to disassemble a three-dimensional garment mentally so that they could cut it out on a two-dimensional piece of fabric. They developed skill through their experiences in sewing and cutting. Mrs. Bowen said that she could cut the sleeve of her client's majorette costume without a pattern because she knew what the sleeve should look like. She said, "You can envision it [the sleeve] … Of course, if you're not used to cutting, you'd never get it cut." By stating "if you're not used to cutting," she implied that her ability to envision the curves and size of the sleeves was developed from her experiences.

After the seamstresses had cut out the pattern pieces from the material, they had to make the flat surface into a three-dimensional surface. They made flat surfaces into cylindrical shapes by attaching the front pieces to the back pieces at the side seams. Because women are not cylindrical in shape, the seamstresses had to fit the garment to the human form. They often inserted darts to achieve this goal. They made darts by sewing an acute angle on a folded piece of fabric. Darts are commonly inserted in the garment's waist area because many women have smaller waists than hips. About 1/2 to 2 inches of material is folded together at the waist. The dart is sewn at an angle so that material is taken in at the waist and tapers out to the hips.

After the dart was sewn, the surface of the material changed from flat to three-dimensional. The dart made the surface of the material appear conelike. The deeper the angle of the dart, the more conelike the surface appeared.

From the second goal, minimizing time and materials to maximize profit, the seamstresses developed skills in estimation and transformational geometry. Karen claimed that her clients would not pay more than $200 for a bridesmaid's dress. Therefore, to make money, the seamstresses had to make the garments quickly and conserve material whenever possible.

Estimation was used by all seamstresses to minimize time. Karen said,

> I'm not really exact ... so I was telling you about Carol [not her name], and we used to laugh `cause she would take her tape measure and say that's 7/16 of an inch. And we'd just laugh. And I will never be that exact. It's not in my nature. I think that's the reason I can make money in sewing.

Using transformational geometry conserved material. For example, Joan used reflection when she flipped over a pattern piece over to squeeze it into an area. She used translation when she moved the pattern pieces closer together while keeping them parallel to the fold of the fabric. Finally, she used rotational symmetry when she turned the pattern 180 degrees.

Although the seamstresses had other goals, such as finding the bias of the fabric, these examples show clearly that they developed mathematics out of the need for accomplishing their goals. Therefore, a characteristic of the seamstresses' mathematics is that it is a necessary and a goal-directed activity.

Thinking

The last aspect that contributes to the uniqueness of the seamstresses' mathematics is their thinking. To characterize their thinking, I knew that it was important to determine what sources contributed to their knowledge and what information constituted their experiences. I found five sources of knowledge that were integrated into the seamstresses' experiences: other experienced seamstresses, directions for patterns, trial and error, other garments, and contexts outside of sewing.

Other women gave the seamstresses dressmaker strategies. Mrs. Bowen learned to sew without patterns by watching her mother. Both Christina and Karen worked for fashion designers. Joan said that she asked her aunt for help.

The experiences with other seamstresses were valuable. Karen said that she learned a lot about how to make garments fit from the designer. In addition, all the seamstresses, except Joan, had the opportunity to observe other seamstresses, for an extended period of time, make garments without using patterns. These experiences probably contributed to their ability to make garments in the same way.

Purchased patterns played an interesting role in the seamstresses' learning. With the exception of Mrs. Bowen, the seamstresses learned how to sew by using patterns. However, these women currently use patterns mostly as guides to cutting out garments instead of relying on them for the rules of construction. They rarely follow the directions step-by-step. Christina said, "Patterns don't tell you enough. One might have something good to say, and the next pattern did it in a more difficult way. So you have to remember the good way to do it."

The seamstresses gained much of their knowledge on their own by experimenting. Some counterintuitive notions were learned through trial and error. Karen said that she learned "the hard way" that she should make the armhole of a sleeve smaller if she wanted the sleeve to be less constraining. Christina said that she also learned this same notion by trial and error. Christina explained that a larger armhole would be more constraining because she would have to cut away fabric to make it larger. Less fabric means that the sleeve will be more constraining.

Most of the seamstresses' reasoning was based on their experiences learned through trial and error. The seamstresses mostly reasoned inductively. For example, the seamstresses found that women had basically the same measurements in certain areas of their bodies regardless of their sizes. Karen said, "Neck to waist is pretty consistent on people, and most people tend to have the same-width shoulders; wrists are pretty consistent; necks are pretty consistent." Similarly, Christina and Joan commented that everyone was basically the same in the shoulder area above the bust. When I told Joan that I did not know that people, regardless of their height, had the same measurements in certain areas, she said, "I didn't know that either until I was sewing; the more you do, the more you learn." The seamstresses drew their conclusions about the shapes of women from their experiences in sewing.

Another source of knowledge was well-made garments. Joan and Karen both learned techniques by examining how other garments were made. Joan said that she learned how to hem suit jacket sleeves with her machine by taking apart a designer jacket . Karen said that she loved to study well-made clothes by "looking inside."

Finally, Christina, Karen, and Joan partly attributed their skills to contexts outside of sewing. Joan believed that her estimation skills came from being mathematics oriented, whereas Christina and Karen associated sewing with art.

Although these were probably not all the sources of knowledge for the seamstresses, they were the ones that I found salient. The information gained was internalized and became known as their experiences. When sewing, the seamstresses chose the strategies or combination of strategies from their experiences that gave them the best result. When their experiences did not help them solve problems, they looked to external sources for knowledge and integrated this new knowledge into their experiences. For example, when Christina explained to me how she learned to cut garments without a pattern, she said, "[You] have to learn the rules and then you can go off on your own and break them." She continued to explain, "[I] read a few books on it…. To me, it's like sculpting." It was evident from her statement that she integrated her personal experiences in art and sewing with the external knowledge found in a book.

Clearly, what constituted the seamstresses' knowledge affected their thinking, decisions, and mathematics. For example, Christina's experiences in art and sewing contributed to her development of designing.

In summary, these women relied on their experiences to solve problems in the workplace. When they realized that their experiences were not adequate, they looked to external sources. Therefore, the seamstresses' thinking can be characterized as a process of reflecting on their experiences, learning new knowledge, and integrating this knowledge with their experiences. Because many of their experiences were similar, their mathematics appeared similar.

The seamstresses saw mathematics in sewing; however, the amount and level of mathematics varied. They did not always identify the mathematics and believed that some of their mathematics was not "valid" or "real" because it was not the kind practiced in school. For example, Christina did not realize that she was constructing a circular arc with her tape measure because she said to me, "I don't know what you're writing down; I'm not doing any math." Karen hesitated to say that her skill of estimation in sewing was a valid form of mathematics until she justified estimation by saying that she would "get down and be more exact" later.

I hypothesize three reasons for the seamstresses not identifying all their uses of mathematics. First, their theoretical awareness of mathematics could be underdeveloped. For example, Mrs. Bowen did not identify the mathematics that she used beyond arithmetic and measurement, probably because of her lack of formal training in mathematics. Mrs. Bowen, unlike the other seamstresses, did not have the opportunity to take a formal course in algebra or geometry because she had left school in the ninth grade.

Second, the seamstresses may have viewed mathematics and sewing as two separate domains. It was evident that Karen viewed mathematics as a male domain.

> *Karen:* I realize now, if I were to take calculus now, I know I would pass it because my approach is different—much more stubborn. I was always frustrated. It always frustrated me because I never understood it. I was never encouraged to pursue a math class. It's a triumph to break a stereotype.
>
> *Sabrina:* Of only men being in math?
>
> *Karen:* Yeah.

I asked her if she pursued sewing because it was a female domain. Although she said yes, she said that she pursued sewing mostly because she enjoyed it.

Karen and Christina may have also viewed mathematics as a domain in which they were unsuccessful. Because they had difficulty with school mathematics, they may not have associated mathematics with their strength—sewing.

Third, there may be mathematics that the seamstresses saw in sewing but failed to describe. Joan and I had similar views of mathematics in sewing probably because we were both mathematics teachers. We both thought of the bias as a direction that was 45 degrees from the grain of the fabric. However, each of us identified mathematics that the other did not see. For example, Joan said that she used angles when she made darts. I never thought that darts were angles until Joan made the comparison. It is likely that the seamstresses and I have identified only a part of the mathematics in sewing.

Because I interviewed and observed only a small group of women, the findings of this study should not be generalized to all women. However, the study does suggest important implications for mathematics education.

The seamstresses were actively engaged with their mathematical symbols and objects: their Cartesian plane of fabric, their symmetrical geometric pattern pieces, and their tape measures. They experienced and developed mathematics through their actions. As mathematics educators, we need to offer students active and tactile experiences in mathematics. They need to be able to touch a plane and to cut out reflections.

Mathematics was necessary for the seamstresses. Therefore, they were motivated to solve problems. As mathematics educators, we need to allow our students to choose mathematical problems that are meaningful to them. Perhaps the students could learn more about their interests through the context of mathematics. The teacher and students could discuss the mathematics that they would use to explore those interests, and the students could appropriately apply the mathematics.

Another way that teachers can help students realize that mathematics is necessary is to have them do projects that require mathematics. For example, when studying statistics, the students could conduct their own surveys and analyze their results. Teachers also can show students examples of how mathematics is necessary and important in life and how it is used at work and at home. We can do this ourselves, or we can invite guest speakers to our classrooms to discuss how they use mathematics with our students .

The seamstresses relied on their experiences to solve problems. If their experiences were not sufficient, they looked to external sources, such as other seamstresses, purchased patterns, and well-made garments, for knowledge and integrated these sources with their experiences. Instead of asking students to solve problems by using our strategies, we need to encourage them to use their own experiences, intuition, and creativity, which may motivate them to explore and pursue mathematics.

Karen viewed mathematics as a context dominated by men and their experiences. One way to eliminate this notion is to integrate into the curriculum examples of mathematics that women have practiced and developed. This curriculum change may empower women in mathematics and enlighten all students about the value of traditional women's work. Mary Harris (1987b, p. 45) stated,

> As learning resources [artifacts of traditional female work], they don't intimidate girls or anyone else familiar with the technology, and they can help reveal just some of the mathematical thinking that goes on among people not reared in generations of failure with standard Western textbooks.

However, if we do bring women's mathematics into the classroom, we must be careful not to devalue their mathematics by implying that girls can do *real* mathematics with their *trivial* sewing. We do not want to foster the idea that women can practice, create, and develop mathematics only in an out-of-school context. Nor do we want this mathematics or the accomplishments of its creators to be trivialized because it does not resemble school mathematics.

Contrary to many people's beliefs, women have been thinking mathematically for a long time. They have created and informally learned their own mathematics through such daily activities as sewing.

Our society needs to value the knowledge that women have developed and realize that they have been creators of, and participants in, mathematics for a long time. Examples of mathematics that have been practiced and developed by women should be integrated into the mathematics curriculum. Broadening our view of mathematics to include the mathematics of women will lead to a richer and more balanced mathematics and a better mathematics for all.

Research on women in mathematics should change. Instead of comparing women with men and determining what women can or cannot do in relation to men, researchers need to study women on their own terms. Women are capable of creating and practicing a complex mathematics, and they are motivated to pursue the study of mathematics in this format. We need to learn more about how women think mathematically in a context created by them and use this knowledge to modify the way that we teach mathematics in school. As a result of our future modifications, I hope that more women will enjoy the field of mathematics and choose to pursue mathematical careers.

REFERENCES

Becker, Joanne R. "Research on Gender and Mathematics Perspectives and New Directions." Paper presented at the annual meeting of the American Educational Research Association, New Orleans, April 1994.

Belenky, Mary F., Blythe M. Clinchy, Nancy R. Goldberger, and Jill M. Tarule. *Women's Ways of Knowing.* New York: Basic Books, 1986.

Bishop, Alan J. "Mathematics Education in Its Cultural Context." *Educational Studies in Mathematics* 19 (1988): 179–91.

Buerk, Dorothy. "The Voices of Women Making Meaning in Mathematics." *Journal of Education* 167, no. 3 (1985): 59–70.

D'Ambrosio, Ubiratan. "Ethnomathematics and Its Place in the History and Pedagogy of Mathematics." *For the Learning of Mathematics* 5, no. 1 (1985): 44–48.

de la Rocha, Olivia. "The Reorganization of Arithmetic Practice in the Kitchen." *Anthropology and Education Quarterly* 16, no. 3 (1985): 193–98.

Fasheh, Munir. "Mathematics, Culture and Authority." *For the Learning of Mathematics* 3, no. 2 (1982): 2–8.

———."Mathematics in a Social Context: Math within Education as Praxis versus Math within Education as Hegemony." In *Schools, Mathematics and Work*, edited by Mary Harris, pp. 57–61. London: Falmer, 1991.

Gerdes, Paulus. "Conditions and Strategies for Emancipatory Mathematics in Underdeveloped Countries." *For the Learning of Mathematics* 5, no. 1 (1985): 15–20.

Harris, Mary. "An Example of Traditional Women's Work as a Mathematics Resource." *For the Learning of Mathematics* 7, no. 3 (1987a): 26–28.

———."Mathematics and Fabrics." *Mathematics Teaching* 120 (1987b): 43–45.

Lave, Jean. "Cognitive Consequences of Traditional Apprenticeship Training in Africa." *Anthropology and Educational Quarterly* 7 (1977): 177–80.

———. "Introduction: Situationally Specific Practice." *Anthropology and Education Quarterly* 16 (1985): 171–76.

Masingila, Joanna O. "Mathematics Practice and Apprenticeship in Carpet Laying: Suggestions for Mathematics Education." Ph.D. diss., Indiana University, 1992. Abstract in *Dissertation Abstracts International* 53 (1992): 1833A.

Millroy, Wendy. *An Ethnographic Study of the Mathematical Ideas of a Group of Carpenters. Journal for Research in Mathematics Education*, Monograph No. 5. Reston, Va.: National Council of Teachers of Mathematics, 1992.

Murtaugh, Michael. "The Practice of Arithmetic by American Grocery Shoppers." *Anthropology and Education Quarterly* 16 (1985): 186–92.

Petitto, Andrea. "Knowledge of Arithmetic among Schooled and Unschooled African Tailors and Cloth-Merchants." (Ph.D. diss., Cornell University, 1979). Abstract in *Dissertation Abstracts International* 39 (1979): 6609A.

Scribner, Sylvia. "Pricing Delivery Tickets: 'School Arithmetic in a Practical Setting'." *Quarterly Newsletter of the Laboratory of Comparative Human Cognition* 6, no. 1 and 2 (1984): 20–25.

Ethnomathematics

A Promising Approach for Developing Mathematical Knowledge among African American Women

8

Gloria F. Gilmer

This chapter speaks directly to the issue of inclusivity with regard to the development of mathematical knowledge among African American women. To this end, the approaches presented here stem from the collective experiences of mathematicians, mathematics educators, and ethnomathematicians, thereby expanding and extending the vision of what mathematics is, who creates it, and in what kind of environment mathematical thinking flourishes for women in general and African American women in particular.

Mathematics is an important human endeavor that has many educational values aside from its technological importance. It offers a vast number of structures, such as numbers, algorithms, shapes, ratios, functions, and data, that are useful in understanding physical realities. It is built on intuitive understandings and agreed-on conventions that are not eternally fixed. Its frontier is covered by many unanswered questions. It encourages settling arguments by evidence and proof. Finally, mathematics demonstrates the importance of subjecting a familiar thing to detailed study and studying something that seems hopelessly intricate (Buck 1965).

To a large extent, educators determine who studies school mathematics and, by extension, who will have careers in mathematics and what the legitimate products of mathematics will be. Yet, every individual is part of a society that has instinctive mathematical knowledge—that is, ways of counting, measuring, relating, classifying, and inferring. Unfortunately, much of this knowledge may be ignored in the formal school mathematics curriculum. Therefore, groups about whom educators are uninformed are bound to receive inequitable treatment in the classroom. The persistence of these inequities often lies in the politics of gender and race.

Against this background, in 1989 I suggested using skits at joint mathematics meetings to dramatize gender inequities—those subtle messages about *who can do mathematics* (Kenschaft and Keith 1991; Wick and Kenschaft 1997). I also suggested including race inequities. These skits were first steps in confronting the inequities between men and women at the professional level within the mathematical community. For many years, these skits were used to raise the awareness level of significant numbers of educators about the nature of gender inequities in the field, yet today these inequities still persist within and across race and gender. This situation suggests that our efforts must be intensified and documented. Therefore, in this chapter, I suggest strategies for providing educational experiences for *all* students in mathematics—especially African American females—oriented toward developing mathematical power.

MATHEMATICS AND ETHNO-MATHEMATICS

In its National Policy Statement 94–95, the American Mathematical Society (AMS) described mathematics as "the study of measurement, forms, patterns, and change which evolved from efforts to describe and understand the natural world" (AMS 1994). Such statements suggest that mathematics is culturally neutral. So, although mathematics educators acknowledged the universality of the truth of mathematical ideas—such as *the sum of the angles of a triangle in a plane is 180 degrees*—this knowledge was divorced from the cultural bases that gave rise to it. For many reasons, such a curriculum has had devastating effects on the representation of African Americans and others in mathematical studies and careers.

Today, mathematics reaches beyond the physical sciences and engineering into medicine, business, the life sciences, and the social sciences (AMS 1994), and mathematics education is regarded as more than a collection of abstract concepts and skills to be mastered. Philosophical arguments about the nature of mathematics focus on what it is that mathematicians actually do (Barton 1985). This new philosophical era brings the mathematical community itself into mathematics in such a way that it is impossible to separate mathematics from this community: its language, preconceptions, values, and experience. Thus, we are witnessing the end of an era in which mathematics was regarded as culturally neutral and the start of an era in which mathematics is acknowledged to be a cultural product. This latter view is strongly supported by ethnomathematicians and has led to the rise of the discipline of *ethnomathematics*.

Ethnomathematics was popularized by Ubiratan D'Ambrosio of Brazil in his keynote address on the relationship between culture and mathematics at the fifth International Congress of Mathematics Education (ICME 5) in Adelaide, Australia, in 1984. To understand ethnomathematics, start with a group bound together by how they use certain mathematical ideas, such as artists, bankers, architects, sports figures, musicians, and seamstresses. Next, examine their language, preconceptions, values, and experience with mathematical ideas—some of which may not be identified as mathematical. Their interaction with these ideas, which may result in certain products within which these ideas are hidden, is what we call their "ethnomathematics."

Researchers in ethnomathematics tend to examine how people learn and use mathematics in distinct cultures and in everyday situations within cultures (Masingila and King 1997). In this context, we may think of *culture* as acquired knowledge transmitted among members of a group. It is shared meaning but not necessarily consensus. It includes taken-for-granted values and beliefs seen in what people do, what they know, and what tools they use (Malloy 1997). In this concept of culture, race is not a proxy for culture, and the "ethno" in ethnomathematics is not a proxy for ethnic.

Since ethnomathematics is oriented to the masses and to the multitude of ways that mathematical ideas are used on a regular basis in a community, the concept expands our understanding of what mathematics is and of who creates it. In ethnomathematics, the focus is on the concepts and techniques actually used by a cultural group rather than on the possible mathematical theories available (Barton 1985). The concepts and techniques are usually learned without formal schooling but are actively transmitted from one generation to another. Through this cultural interaction, common mathematical knowledge develops among the adults and children who belong to the same cultural group (Gilmer 1985). Later in this chapter, this phenomenon is illustrated in the ethnomathematics of hair braiders.

An ethnomathematics curriculum would develop from the learners' surroundings and move seamlessly into the school as the process of inducting young people into the mathematical aspects of their culture. A mathematics curriculum oriented to the ethnomathematics of the learners' culture would respond to the needs of the increasing numbers of students who feel like failures for not understanding something that few of them will ever use but without which there is the perception of a bleak future.

If we acknowledge that mathematics is not culture-free, then mathematics educators might be transmitting the values of a single culture while teaching children of many cultures in the same classroom. What are the consequences of mathematics learning for the children whose cultural experiences are being ignored? One consequence might be mass disaffection with the subject owing to cultural conflicts or school failure. An example of this was given by Malloy (1997) in relating the following test item whose solution was judged incorrect because the test designers and the students made different assumptions.

> It costs $1.50 each way to ride the bus between home and work. A weekly bus pass is $16. Which is the better deal, paying the daily fare or buying the weekly pass?

On the one hand, test designers assumed that only one person would use the bus pass, the pass would not be used on weekends, and the person would have only one job. On the other hand, many African American students assumed that three or four people might use the same pass at different times of the day or on weekends or that if only one person used the pass, he or she could have two jobs. Since situational mathematics is almost always culturally based, in multicultural settings care must be taken to include cultural assumptions in the statement of the problem.

CULTURAL INFLUENCES ON MATHEMATICS TEACHING IN OUR CONTEMPORARY SOCIETY

To incorporate students' interests into the mathematics curriculum, one might first have students explore activities observed in their own surroundings. Bishop (1988), Gilmer (1990), and Gerdes (1997) suggest where one might look in the learner's environment for clues to mathematical behavior—the products they design and how they count, measure, locate, play, and explain. By exploring how different "ethno groups" in their community conceptualize, code, and symbolize mathematical ideas found in these universally significant activities, the curriculum becomes relevant to classroom learners in a very natural way and is simultaneously multiculturalized (Gilmer 1990).

An example of this approach is my study of hairstyles in African American communities (Gilmer 1998). The idea of the study was to determine what the hair-braiding and hair-weaving enterprise can contribute to mathematics teaching and learning and to find out what mathematics can contribute to the enterprise. This study led me to observe African American hairstylists at work. Hairstylists were interviewed along with their customers. One case revealed rectangular tessellations of the scalp using a pattern that started at the nape of the neck and increased by one rectangle at each successive row leading away from the neck. The pattern is illustrated figure 8.1a.

In figure 8.1b, the dots mark places from which braids emanate—roughly at the point of intersection of the diagonals of the rectangles. When asked why she used this pattern, the hairstylist said that it was a space-filling pattern that hid

THE ETHNO-MATHEMATICS OF HAIRSTYLES IN AFRICAN AMERICAN COMMUNITIES

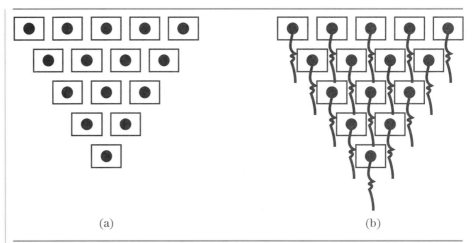

(a)   (b)

Fig. 8.1. Patterns of hairbraiding

the side of the rectangle at the previous level where the hair was parted on the scalp. On examining this space-filling pattern, I realized that the number of braids might be a more equitable pricing unit for hair braiding than a flat rate, since hair braiding is labor intensive. The hairstylists interviewed in the study, however, had no idea of the number of braids completed. This number could be determined mathematically with the following simple formula:

$$S = 2 + 3 + 4 + ... + n = n(n + 1)/2,$$

where S is the number of braids and n is one more than the number of rows. Hence, for the four rows above, the total number of braids is

$$S = 5 \bullet 6/2 = 30/2 = 15$$

(When using microbraids in hairstyles, where upward of 700 braids may be involved, the hairstylist would find a formula for unit pricing by braids especially useful in determining the cost for such styles.)

In another example, the customer's scalp was tessellated with triangles. Hair strands within the triangle were brought to the center of the circle inscribed in the triangle. At that position, braiding commenced (see fig. 8.2). Braids so formed were said to be less likely to swing with head movements than braids formed by bringing hair strands in the triangle to a vertex of the triangle that points floorward (see fig. 8.2b).

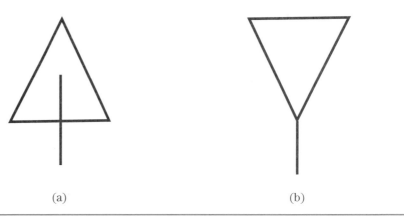

(a)   (b)

Fig. 8.2

The topic of hairstyles in mathematics is natural when we consider what cultural groups design. Many styles involve interesting geometrical designs on the scalp, such as spirals and circles (Sagay 1983). Many such styles are cross-gender and cross-cultural. In addition, this topic offers insight into some cultural values that form the basis of hair-braiding and hair-weaving traditions in African American communities. For generations, African Americans were told that "nappy" hair was bad and were made to feel that the only way to attain "good" hair was to straighten it. Strong chemicals and heat treatments used to straighten the hair often resulted in damaged, unhealthy hair that would not grow. The customers interviewed felt good about having a beautiful hairstyle without altering the natural texture of their hair. Beyond beauty, the hair-braiding enterprise is also an important source of income for African Americans. One stylist said that at the age of eleven, she was the neighborhood braider and could always earn money. The concepts of time spent, price setting, and costs of supplies and equipment are all important sources of mathematical problems for the classroom derived from this topic (Gilmer 1998).

Listen to the Students

Our classroom cultures are our most accessible and relevant sources of information for curriculum development. Effective classrooms for African Americans encourage high levels of peer interaction, group decision making, expressiveness through appropriate wait time, physical closeness, acknowledgment, feedback, probing, and listening (Malloy 1997). Teachers should involve themselves in the experiences of their students by exploring the students' community and extending community activities into the classroom practice. Alan Bishop (1988) contends that children are creating their own culture, not just managing the culture of their ancestors. Hence, mathematics education should be oriented more to the present and future than to the past. We must allow students to teach us about the culture they are creating, just as we teach them about the foundations on which their culture is being built.

A proponent of this view is David Henderson. For more than two decades, Henderson has taught junior- and senior-level geometry courses for mathematics majors and future teachers at Cornell University. Henderson's teaching style is similar to that of the late R. H. Bing at the University of Wisconsin, under whom the author also studied. Bing taught without lectures or textbooks. Bing listened to his students and encouraged them to express their understandings and reasoning in their own words. By using this same approach, Henderson eventually discovered that he was learning from his students. In his paper, "I Learn Mathematics from My Students," he gives examples of new theorems and proofs—shown to him by his students—none of which had appeared in print (Henderson 1996).

The data in table 8.1 show that white women, men of color, and blacks all have higher "percents who showed me new mathematics" than white men. Henderson says this is significant because each of these groups is underrepresented in mathematics in the United States. He concludes that he must listen particularly carefully to the meanings and proofs expressed by persons of color and women because there is much that they see that he does not see! Table 8.1 shows that on a percentage basis relatively fewer women of color showed Henderson new mathematics. This, says Henderson, may indicate that he may need to listen more carefully to women of color. He acknowledges that hearing someone else's proof may be difficult and require considerable effort and patience on his part. Henderson concludes, "Perhaps it is true that women and persons of color are underrepresented in mathematics because they are not being well listened to by those of us already in mathematics."

STRATEGIES FOR INCLUSION

Table 8.1

	Number of Students	Number Who Showed Me New Mathematics	Percent Who Showed Me New Mathematics
All students	178	56	31
White men	85	25	29
White women	58	21	36
Women of color	22	5	23
Men of color	13	5	38
All blacks	10	4	40

Promote Exploration

To incorporate students' interests into the mathematics curriculum, one might first have students explore activities observed in their own surroundings. Both Bishop and Gerdes suggest where one might look in the learner's environment for clues to mathematical behavior—the products they design, how they count, measure, locate, play, and explain (Bishop 1988; Gilmer 1990; Gerdes 1997).

Designing—This activity concerns all objects and artifacts that cultures create for various purposes, from home life and adornment to warfare. The designed objects often serve as models for the construction of other objects and are sources of important mathematical ideas, such as shape, size, scale, ratio, proportion, symmetry, and many other geometrical concepts.

Counting—This activity relates to what, how, and why people count and includes a variety of counting systems developed by indigenous groups.

Measuring—This activity is concerned with comparing, ordering, and valuing. Precision measurement and systems of measuring units develop in relation to what the society values. An example is how housing costs are determined.

Locating—This activity relates to finding one's way around, traveling without getting lost, and relating objects to each other. All societies have developed different ways to code and symbolize their spatial environment—witness the highway system in the United States. The resulting conceptualizations and explanations, however, may differ from culture to culture.

Playing—All cultures play. This activity connects to mathematics when it is formalized into the notion of games. The development of games involves behaviors that are rule governed in a manner similar to the rule-governed criteria of mathematics. Examples of such rule-governed behavior are the estimation of angles and distances required in basketball or the logic applied to moves in the game of chess.

Explaining—This activity exposes connections between apparently diverse phenomena, allowing for a kind of unity from which mathematical proof is derived.

By exploring how different "ethno groups" in the community conceptualize, code, and symbolize the mathematical ideas found in these universally significant activities, the curriculum becomes relevant to classroom learners in a very natural way and is simultaneously multiculturalized (Gilmer 1990).

Explain What to Look For

Mathematical power involves the ability to discern and investigate through a variety of mathematical methods the mathematical relationships observed in patterns and structures in one's own surroundings. Encourage students to search first for patterns in the activity studied and next for mathematical relations embedded in those patterns. This is best done by exploring special cases in a systematic way. From this investigation, patterns may emerge that will suggest ideas for proceeding with the problem (Larson 1983).

Teach from the Students' Vantage Point

A study of the learning styles of Canadian women in the trades and technologies might guide the teaching and learning of African American females in similar fields and in mathematics as well (Brooks 1986; Gilmer 1989). Women reported that they learn best if presented with an overview of the material during which they can (*a*) relate it to themselves, (*b*) see a demonstration, and (*c*) shift back and forth between application and discussion. They observed that their learning takes the following sequence:

1. Understand the value of what I need to learn.
2. Hear what I need to learn.
3. See what I need to learn.
4. Talk about what I need to learn.
5. Do what I need to learn.

Learners in this study were classified on the basis of three faculties: mental, relational or emotional, and physical. *Mentally centered learners* were said to focus on ideas and rely on articulation skills to convey what is learned. For them, instruction need not be relevant to their daily experience. Mentally centered instructors are said to rely on verbal skills, lecture, and analysis in teaching. At Fanshawe College, the site of the Canadian study, there were no mentally centered learners among the students and instructors assessed. Yet, the study notes that mentally centered instructional forms have been adopted widely by teachers, most of whom are not mentally centered learners. This could have serious implications for the quality of teaching.

Ninety-three percent of the students assessed were *relational learners*. Feelings are the focus of learning in this group and play a role in how fast students learn. At first, these students may work to satisfy the instructor. Later, they work to satisfy personal goals. Their learning is prompted by linking new material to things already known. This allows them to draw on discreet or intuitive knowledge. They rely on verbalization to convey ideas and are best served if they move into application or hands-on work after some, but not all, information has been presented. Application makes the learning concrete for relational learners.

Seven percent of the students assessed were *physical learners*. Members of this group prefer brief, orderly, concise presentations of material or directions. They apply their learning methodically. They learn by doing and by being given enough time to stick to a task until it is completed. They learn through repetition and work well on details. A sense of belonging within the group is important to these learners. Such learners may be good at mathematics and science. The majority of Japanese and Chinese learners are said to be physically centered.

These three types of learners differ mainly in their initial processing of information. The study concluded that relational and physically centered learners in nontraditional programs could be well served by employing interactive instructional styles and hands-on applications early in the learning process, with movement back and forth among theory, discussion, and application. This learning process might also serve African American females well in their study of mathematics!

CHALLENGES TO INCLUSION

People who believe they gain status from their children's performing well in school do not like the idea that the other kids' performance might be raised to the level of their own kids (Kohn 1998). According to Kohn, they are not concerned that all children learn, just that their children learn. They do not see school as a place for learning but, instead, as a place for accumulating credentials. Kohn claims that often such people are white, middle-class parents of high-achieving students. These parents might be some of the community's most outspoken and influential members. They have learned how to work within the law and how to use the law skillfully.

The people mentioned above are typically involved in three types of controversies: (1) the type of instruction, (2) student placements, and (3) the awards systems. Instead of advocating instruction leading to active discovery and problem solving by a community of learners, these parents favor a return to a skill-and-drill mathematics curriculum and an individualistic competitive credentialing model of school that may boost their children's SAT scores, thereby enhancing their chances for acceptance into the most elite colleges. With their substantial political power, they fight efforts to create more heterogeneous and inclusive classrooms—preferring instead ability grouping, gifted-and-talented programs, honors courses, and a tracking system that keeps virtually every child of color out of advanced classes. In San Diego, California, they vigorously opposed a program to provide underachieving students with support that would help them succeed in higher-level courses.

Finally, these parents favor practices that distinguish one student from another, such as letter grades, weighted grades, honor rolls, and class ranks, so that only a few will be recognized at awards ceremonies. In Buffalo, New York, for example, Kohn says parents of honor students squashed an attempt to replace letter grades with standards-based progress reports. Arguably, their agenda has little to do with meeting children's needs. What remains unrecognized by some privileged parents is the evidence that enrichments such as hands-on learning, student-designed projects, computers, and field trips, and not selective placements, produce better results for their children.

In commenting on the belief that the elite 20 percent do not care about the remaining 80 percent as long as life is good for them and their children, Thacher (1997) noted that Asian countries focus less on skills in early elementary education than on social collaboration. Teachers in Japan and China put primary emphasis on the importance of young children's learning to work together across social gaps. These schools, for the most part, avoid tracking in the early years. The heavy emphasis on skills and academic results is delayed until the more advanced stages of education.

Chinese immigrants to this country in the 1930s brought a balanced emphasis on hard work and socialization. They held a deep conviction that childhood is an end in itself. Thacher claims that in the face of a materialistic culture, these immigrants fostered the values of simplicity and friendliness in their schools—"as radical as they must have sounded to affluent suburban parents." In addition, the students who attended these schools have been able to cling to these values ever since. Such values are needed today to reverse numerous obstacles to the development of an inclusive community of learners.

The educational philosophy of Clarence Stephens—one of my undergraduate mentors at historically black Morgan State University in Baltimore, Maryland— had a profound impact on my own career. He challenged his students to learn how to learn, for he believed we would encounter some teachers who could not teach us. He cautioned teachers in the mathematics department to suspend disbelief with respect to students whose past records have been undistinguished and to inspire all students to achieve at a higher level. He claimed that the secret to getting students to succeed is to keep up morale. Finally, he felt that in an effective learning community, there must be total support for students and teachers alike. In order to follow Stephens's concept of an effective learning community, the school community must develop a strategy to encourage, challenge, and train its members to know one another. In particular, the school community must have a plan to assure equitable support for marginalized students. Parents typically lacking in wealth, self-confidence, or political savvy must be provided with knowledge and skills that will make them more effective advocates for themselves and their children.

Empowering students involves considering their beliefs about themselves and their learning environment and enlisting their active participation in cocreating their learning environment. The instructional approaches of both Stephens and Henderson empower students. Specifically, these approaches allow everyone an opportunity to understand and respect one another's knowledge and ways of knowing. In this way, students develop self-confidence in mathematics. Two important components of self-confidence are (1) valuing mathematics as being important for success in life and (2) perceiving that one has the ability to be successful at mathematics (Fleener, DuPree, and Craven 1997). The mathematical power needed to develop self-confidence includes the ability to discern and investigate mathematical relationships observed in patterns and structures in one's own surroundings using a variety of mathematical methods. In this way, mathematics may be viewed as the process of inducting young people into the mathematical aspects of their culture.

American Mathematical Society. *National Policy Statement 94–95*. Washington, D.C.: American Mathematical Society, 1994.

Barton, Bill. "Ethnomathematics and Curriculum Change." Unpublished paper, December 1985. (Available from the author at Department of Mathematics, University of Auckland, P.B. 92019, Auckland, New Zealand)

Bishop, Alan. "Mathematics Education in Its Cultural Context." *Educational Studies in Mathematics* 19 (1988): 179–91.

Brooks, Carol. *Working with Female Relational Learners in Technology and Trades Training*. London, Ont.: Fanshawe College, 1986.

Buck, R. C. "Goals for Mathematics Instruction." *American Mathematical Monthly* 72 (November 1965): 949–56.

Fleener, M. Jayne, Gloria Nan DuPree, and Lary D. Craven. "Exploring and Changing Visions of Mathematics Teaching and Learning: What Do Students Think?" *Mathematics Teaching in the Middle School* 3 (September 1997): 40–43.

Gerdes, Paulus. "On Culture, Geometrical Thinking and Mathematics Education." In *Ethnomathematics: Challenging Eurocentrism in Mathematics Education*, edited by Arthur B. Powell and Marilyn Frankenstein. Albany, N.Y.: State University of New York Press, 1997.

Gilmer, Gloria F. "Sociocultural Influences on Learning." In *American Perspectives on the Fifth International Congress on Mathematical Education (ICME 5)*, edited by Warren Page. Washington, D.C.: Mathematical Association of America, 1985.

————. "Making Mathematics Work for African Americans from a Practitioner's Perspective." In *Making Mathematics Work for Minorities: Compendium of Papers Prepared for the Regional Workshops*, pp. 100–104. Washington, D.C.: Mathematical Sciences Education Board, 1989.

————. "An Ethnomath Approach to Curriculum Development." *International Study Group on Ethnomathematics Newsletter* 5 (May 1990): 4–5.

Gilmer, Gloria F., and Mary Porter. "Hairstyles Talk a Hit at NCTM." *International Study Group on Ethnomathematics Newsletter* 13 (May 1998): 5–6.

Gilmer, Gloria F., and Scott W. Williams. "An Interview with Clarence Stephens." *UME Trends* 2 (March 1990).

Henderson, David W. "I Learn Mathematics from My Students." *For the Learning of Mathematics* 16 (June 1996): 46–52.

Kenschaft, Patricia, and Sandra Keith, eds. *Winning Women into Mathematics*. Washington, D.C.: Mathematical Association of America, 1991.

Kohn, Alfie. "Only My Kid: How Privileged Parents Undermine School Reform." *Phi Delta Kappan* 79 (April 1998): 569–77.

Larson, Loren C. *Problem Solving through Problems*. New York: Springer-Verlag, 1983.

Malloy, Carol E. "Including African American Students in the Mathematics Community." In *Multicultural and Gender Equity in the Mathematics Classroom*, 1997 Yearbook of the National Council of Teachers of Mathematics, edited by Janet Trentacosta, pp. 23–33. Reston, Va.: National Council of Teachers of Mathematics, 1997.

Masingila, Joanna O., and K. Jamie King. "Using Ethnomathematics as a Classroom Tool." In *Multicultural and Gender Equity in the Mathematics Classroom*, 1997 Yearbook of the National Council of Teachers of Mathematics, edited by Janet Trentacosta, pp. 115–20. Reston, Va.: National Council of Teachers of Mathematics, 1997.

Sagay, Esi. *African Hairstyles*. Portsmouth, N.H.: Heinemann Educational Books, 1983.

Thacher, Nicholas S. "A Headmaster Reflects: Notes from an International Conference of Elementary School Principals." *Education Week*, 3 December 1997, p. 39.

Wick, Catherine Anne, and Patricia Clark Kenschaft. "Microinequity Skits: Generating Conversation about Gender Issues." In *Multicultural and Gender Equity in the Mathematics Classroom*, 1997 Yearbook of the National Council of Teachers of Mathematics, edited by Janet Trentacosta, pp. 209–13. Reston, Va.: National Council of Teachers of Mathematics, 1997.

One White Male's Reflection on Participating in Experiences Related to Gender Equity

An Interview with Dale Oliver

9

Dale Oliver

Judith E. Jacobs

D ale Oliver has participated in several experiences aimed at increasing awareness of equity issues. He sat down with Judith Jacobs to discuss part of his personal journey related to gender-equity issues. This interview is presented so that the readers of this book could consider one white male's reaction to such experiences.

Judith:

Dale, in several professional-development projects with which you work, you require the participants to engage in experiences related to equity, knowing that they may cause them some disequilibrium. Why do you feel these are so important?

Dale:

Lack of equity, related to both gender and culture in the field of mathematics teaching, is an issue about which many in the mathematical community are unaware or uninformed. There is a pervasive feeling that the field of mathematics and its teaching are somehow isolated as a science from these "social" issues. At the very least, experiences related to equity are important to raise awareness in, and provide information, to the mathematics community. In the best situation, such experiences can transform the teaching of mathematics and improve educational experiences for all students.

As for disequilibrium, the professional-development models of my projects depend on creating some disequilibrium for participants—and then providing some significant time for personal and group reflection—so that participants are moved to examine their underlying beliefs about what they teach and how they teach. Even so, I am comfortable dealing with disequilibrium on an intellectual level, but not so on the more emotional level that dominates experiences related to equity. Initially I was reluctant and fearful to include the emotional strain of such experiences in the professional-development programs that were my responsibility. I felt that perhaps we in the mathematics community had no business delving into such personal and emotional areas with our colleagues.

Experts had told me that wrestling with personal and group views of equity is an essential component of professional development for teachers of mathematics. It took personal experience for me to believe that this approach was necessary.

My initial experiences in equity sessions were personally traumatic. I am a good teacher. I work hard to be sensitive to the needs of my students, and I work hard to enable each of my students to build knowledge, understanding, and skills that are useful and important. Even so, as sensitive and earnest as I believe I am, I was shocked at the lack of my own awareness of the different educational experiences, social values, and ways of thinking that are represented in my college classrooms. In spite of my use of inclusive language, my use of a variety of motivational examples, and my attention to individual students, I still had been measuring mathematical power for my students by how close they came to my way of thinking and feeling about mathematics. I was unaware of how my attitude alienated those who were not like me and affirmed those who were like me.

I became more convinced of the value of sessions on equity as I interacted with other colleagues who were having similar reactions. There are many dedicated and talented faculty who have their students' best interests in mind and yet do not have the expertise or awareness to adequately examine issues of equity as they relate to teaching mathematics. So, from the standpoint of individual professional development, sessions on equity are one of the foundational experiences needed for examination of what and how one teaches.

On the group level, colleagues who share common experiences of disequilibrium (even from very different perspectives) have a basis for discussing ways to improve the situation. Ultimately one hopes that out of such discussions come improvements in instruction that are better for those students who continue to be marginalized by current practice.

Judith:

Some of these experiences provided theoretical models for why gender is an issue in mathematics education. These often present research results or suggestions for interventions at the classroom level. One such research result describes the different ways that females and males respond to success and failure. A controversial intervention in the classroom is the use of feminist pedagogy, in which the teacher and the students share authority for deciding whether something is correct or incorrect. What was your response to these?

Dale:

As in all areas of study, the most valuable theoretical models are those that are supported by research findings. Even so, for most teaching practitioners, an awareness of theory and research is not enough. Without suggestions for practical applications or strategies for intervention that are linked directly to teaching mathematics, the sessions on equity can leave people feeling confused, frustrated, and powerless. Untested theoretical models or untested suggestions for intervention are routinely met with skepticism. However, the theoretical models and research findings play an important role in debunking the myth that "good teaching practices" alone produce mathematics classrooms that are free of bias.

Judith:

What made these experiences comfortable for you?

Dale:

I expected you to ask, "What made these experiences uncomfortable for you?" I'm sure that we will get to that later. Most of these experiences have been uncomfortable for me. However, sessions on equity are more "comfortable" when the facilitator—

- establishes ground rules for mutual respect, including limiting comments about what people said to the session itself;
- warns the participants about potentially difficult or sensitive topics, giving permission for people to feel anger, guilt, and grief;
- acknowledges the good will and intentions of all in the room (otherwise they would not be in the room);
- keeps the discussion expressive and emotive, but not confrontational;
- organizes the discussion so that all who want to may speak and no individual or group dominates the conversation.

Comfort also comes when a group of colleagues has been through the initial stages of listening, awareness, pain, and grief and has moved on to embrace a spirit of reconciliation and partnership in seeking practical solutions and strategies for valuing all people in the mathematics classroom.

Judith:

We have jointly participated in an experience in which the leader was able to tap into some deep feelings among participants regarding their own sense of "otherness" or discrimination or their pain and outrage at the inequities targeted at others. What was your reaction to that experience? Reflect on the first time you experienced or witnessed discrimination. What did you feel and how did you react?

Dale:

In hearing about others' experiences of discrimination, I was most startled by the depth and breadth of pain that occurred recently in the contexts I view as safe and open. For example, hearing about ongoing discrimination in public education astounds me. I believe that I represent an optimistic (and perhaps naive) segment of the population who believes (or believed) that civil rights issues were settled during the 1960s. At that time, I was attending a suburban elementary school largely untouched by the turmoil of the era. Moreover, I do not have any personal experiences of discrimination from my past that were significant enough to remember in detail, and yet I have been in rooms with others for whom there are many serious instances. The first time I heard such stories, my reaction was part shock and part disbelief. I thought, "How could people experience such discrimination and not do anything about it? Why wasn't it fixed? Is there more to this session than bemoaning the fact that reality is not matching our dreams?"

My own observations of discrimination that are personally painful— in terms of both the discrimination and my reaction to the discrimination—have been gender-related. Two women who are very close to me have both struggled with lower wages and fewer opportunities in their professions than comparable male colleagues. I have been angry at the lack of fairness and have wanted to set the offenders straight (as if my "fixing" the situation for them was my prerogative as a male), but I took no action. In each situation, I attributed the discrimination to the pervasive culture of male-dominated professions in which change had occurred, but not as quickly as desired. I suggested that we be patient.

My wife experienced significant pressure to be "more like a man" when she was in graduate school in architecture. She was criticized for showing emotion during oral defense of her work. Her professors implied that she had to do better than her male colleagues to receive the same measure of success. I was furious! Her professors said that they were doing this for her own good—to get her "toughened up" for the real world. I let it go.

Judith:

Dale, how do you think that you would respond today to an analogous situation? Say a student comes to you and tells you that another mathematics professor is not treating men and women students the same. How would you feel? What would you do?

Dale:

I wish that this were a hypothetical question. My reaction in such a situation is a mix of anger and apprehension. I am angry that the student has had to focus valuable time in the educational process to deal with this issue. I am angry at the potential harm done to the student. Perhaps the student now feels less valued intellectually and personally. Perhaps the student feels less capable of future success because of who she or he is. In any event, the student who has come to make the complaint has been affected significantly enough to go to another person for direction.

The apprehension I feel is connected to the inevitable confrontation with my colleague. How can I approach the topic discreetly enough to prevent my colleague from becoming defensive? How can I approach the topic forcefully enough to generate reflection and change for the better? How can I be open enough to hear my colleague's perceptions and reflections and to assist in making positive changes that may push me to reexamine what I do professionally?

Although I do not like confrontation, I (or any of us) cannot afford to let these things go. In such situations, privately confronting the offending colleague (while preserving the confidentiality of the student who brought the complaint) is a must. Not only must I be ready to discuss the complaint, but I must be willing to continue the dialogue, including frank discussions of underlying beliefs and the practices that support those beliefs.

To date I have been more successful at keeping the confrontations private and nonoffending than I have been at stimulating reflective analysis and positive change of behavior. My belief about what I ought to do is still miles ahead of what I practice.

Judith:

Some activities in your programs have resulted in some participants feeling blamed for causing or perpetuating the inequities in mathematics education and becoming defensive. What was your initial reaction to some of these more affective activities?

Dale:

A defensiveness always wells up inside of me whenever I am in a session focused on gender or cultural equity. Because I am a member of the traditionally dominant gender and culture, I feel guilt by association. I cannot avoid these selfish feelings, even though I know that they take away from my attending to the real pain that others have experienced. I know that such sessions are not designed to place blame on the white males in the room or on anyone else. I also know that reconciliation and growth cannot happen without discussing and airing painful experiences of prejudice and bias. So, the uncomfortableness will (should) remain as a necessary part of the process—at least initially. The process stagnates when the feelings of defensiveness become a focus of the group discussion.

Many white males do recognize their privileged history and recognize the injustice that continues for women and people of color. We need to be allowed to express our sorrow and our feelings of guilt, but then move on to participate in reconciliation. We need guidance in how to participate in righting the wrongs around us as we continue to work within our profession.

Judith:

As you reexperienced these activities or experienced similar ones, did your reaction change?

Dale:

With each experience there still is a period of grief about past and ongoing discrimination and inequities. This is an unavoidable and important part connected to the human experience. What has changed with time and exposure are my ability to listen to what others have experienced (instead of focusing on my reaction to their experience) and my desire to work collaboratively with others in making improvements related to gender equity. I have also grown less patient with colleagues who refuse to listen to the pain of others or who dismiss such pain as irrelevant to the teaching of mathematics.

Judith:

How would you describe your current state with respect to gender issues in mathematics?

Dale:

I feel much more aware of issues related to gender equity in my own teaching and much more empowered to apply expert counsel and advice in my own classroom. I have been assisted by important books, such as *Women's Ways of Knowing* (Belenky et al. 1986), *Toward Gender Equity in the Classroom* (Streitmatter 1994), and *New Directions for Equity in Mathematics Education* (Secada, Fennema, and Adajian 1995). I have also been assisted by many colleagues who continue to be willing to share what they do in their own classrooms. Using these resources, I have been able to lead sessions on gender equity directly with my preservice teachers.

I must admit that leading sessions on gender equity is some of the most exhausting work that I have ever done. Even so, I have seen such sessions turn out to be pivotal in the professional development of new teachers. I have had some students claim that it was the "emotional disequilibrium" of the equity sessions that got them to see themselves as teachers of students rather than as mere presenters of mathematics.

Judith:

Do you find that you had the same reaction to equity experiences related to gender and to those related to race and ethnicity? If not, how did your reactions differ?

Dale:

My reactions do not differ significantly, except in the initial stages in which I have more personal experiences with gender equity. However, I do not want to suggest that equity experiences related to gender should be lumped together with those related to race and ethnicity. The issues involved are not all the same. There are unique aspects of awareness, information, and intervention that deserve full attention for each.

Judith:

Do you have any other insights to share with those of us who conduct gender-equity workshops?

Dale:

In my estimation, participants who are teachers of mathematics generally do not associate their work with social issues. The sessions on gender and race and culture must be tied to the specific context of mathematics teaching. There is a sense in the mathematics community that fairness rules in mathematics. How can the teaching of mathematics be biased? Mathematics makes no judgments about a person's character, motivation, or home life; it is just mathematics. Belief in this kind of inherent fairness in the teaching of mathematics must be examined in a way that moves people beyond defensiveness to an honest analysis of teaching practices and potential interventions.

Leading colleagues to face the realities of gender inequity is thankless (and exhausting) work, but it is work that has to be done. I have great respect for the work that you and others do. Remember that when we who are participating in your workshops are ready, we need concrete advice. Too much disconnected material confuses the issues that can most likely be addressed by the people in the workshops.

Judith:

Dale, thank you for sharing your reactions and insights into how one white male has grown through experiencing several activities on diversity, particularly those related to gender. Your comments will be a reminder that individuals need to experience these types of activities more than once and over time because they need time to distance themselves from their immediate reactions and to assimilate the ideas and feelings the experiences prompted.

REFERENCES

Belenky, Mary, Blythe Clinchy, Nancy Goldberger, and Jill Tarule. *Women's Ways of Knowing: The Development of Self, Voice, and Mind.* New York: Basic Books, 1986.

Secada, Walter G., Elizabeth Fennema, and Lisa Byrd Adajian. *New Directions for Equity in Mathematics Education.* New York: Cambridge University Press, 1995.

Streitmatter, Janice. *Toward Gender Equity in the Classroom.* New York: State University of New York Press, 1994.

ABOUT THE INTERVIEWEE

Dale Oliver is an associate professor of mathematics at Humboldt State University in Northern California. There he teaches a broad range of undergraduate mathematics courses as well as methods courses for prospective secondary school mathematics teachers. His degree is in mathematics, and his experience includes teaching high school mathematics in the Detroit Public Schools.

Since 1992 he has been the codirector (with Phyllis Chinn) of the National Science Foundation–funded Project PROMPT (Professors Rethinking Options in Mathematics for Prospective Teachers), a college faculty-development project for improving the mathematical preparation of future elementary school teachers. He is also the codirector of the Redwood Area Mathematics Project, a state-funded faculty-development project for grades K–12 teachers. Both projects present professional development through intensive residential summer workshops, which include sessions on issues of equity in mathematics education.

Alice and Rachel

Teacher-Leaders of Color

10

Ana Becerra

An unspoken space exists between the understandings of issues of equity held by white women and by women of color. Although we seem to agree on the goal to eliminate gender bias in mathematics education, we enter the arena carrying with us very different baggage, in the form of our prior experiences. In this chapter I talk about Alice and Rachel, two teacher-leaders of color. Although they have never met, I have woven together their stories to help explain what it is like for women of color when they decide to take leadership on issues of equity in mathematics education.

A comprehensive study of female administrative aspirants found that competition exists between minority and white females (Banks 1995). This finding raises an important question about the extent to which white women and women of color will support and cooperate with each other. The reality of competition in the workplace may result in the perception that the advancement of one group of females may hinder the advancement of other groups. Statistics from the U.S. Department of Labor clearly indicate that the dual status of being minority and female results in lower earning power for black women—thus disproving the myth that black women get top jobs simply because they are minority women. Although women and minorities encounter barriers to leadership positions, minority women confront both gender and racial barriers (Edson 1987; McCarthy 1993).

There may be a connection between the small number of studies on minority women in school administration and the low number of minority women actually in school administration. Women of color were almost completely absent from the decades-long scientific research on leadership until the late 1970s. Research on and by women of color in educational leadership continues to be scant. The lack of research on women of color in leadership was not viewed as problematic because race was not considered a difference of consequence.

Generally, issues of race and gender have been explored as independent processes. Considerably more research is available on women than on people of color in educational leadership. This tendency ignores women of color as an integrated whole and instead presents them as fragments. Yet for women of color, experiences of discrimination are based on two factors: race and sex. Researchers seemed to assume that their findings could be applied without regard for race.

Also lacking in the current research are ways that administrators, professional developers, and policymakers can encourage and support racially diverse teacher leadership capable of bringing alternative perspectives to the development and implementation of policies that affect underrepresented students. The lack of acceptance of, and support for, a variety of forms of leadership is an impediment to the development of educational leadership among women of color.

ALICE AND RACHEL: TWO TEACHERS OF COLOR

I know Alice and Rachel from their participation in a number of mathematics equity education projects. They are both classroom teachers and leaders at their Southern California schools. Alice is an African American mathematics teacher at a junior high school; Rachel is a Chicana teaching bilingual kindergarten in a barrio elementary school.

Growing up in South Central Los Angeles, Alice echoed the struggles faced by many young people learning to survive in high-poverty communities. Although she dropped out of high school at fifteen, she always knew that she was smart. Alice held onto the dream that despite the obstacles, one day she would attend college. She saw this as her only way out of the struggles she witnessed and participated in daily. Hers was a world of inequities and human tragedy that few in the mainstream culture will ever know. Even through her serious involvement with drugs and the birth of her son before she was eighteen years old, Alice held on to her plan for finding a way to a better life. Alice credits part of the realization of her plan to become a teacher to the support of the man she married.

Rachel, one of eight children, grew up in an agricultural community. She has found her role in leadership to be "scary." Although she does not find it difficult to "get up sometimes and talk," mathematics is an area in which she does not feel a sense of expertise. Like many bilingual teachers, she has focused her energies in areas of reading or language arts. As a college student, Rachel had no intention of teaching until she worked at a prison for youthful offenders:

> It was just a nightmare, a living hell. When you go there, you can just see that the inmate population is so disproportionately minority. I came into teaching with a very clear desire to make a difference for the kids so that they wouldn't end up over there. All through school I had no desire to be a teacher. None whatsoever. I just never want to see any of these kids over there, and I see the way society is laid out. There's not a lot of options for a lot of people of color. There's not a lot of options in your life really if you do not have an education. It should be society's big push, to give these children everything we can now because it's going to benefit us later. And if we don't, the opposite is going to happen. That's what's happening. When I am working [with students in the classroom], I am thinking of those kids. I always remember when I was leaving on my last day. A boy was yelling at me through the bars on the window as I walked by, "Don't forget about us. Don't forget about us here." It's a very emotional thing to remember (crying) because I do [remember].

Taking Leadership

Enthusiastic and ambitious, Alice is always thinking about ways to continue to grow and learn. She has eagerly enrolled "in every workshop. You name it, I'm there"—even when she must pay her own way. She believes that by putting more into her own development, she will have more to give to her work. Alice admits that initially she didn't view herself as a leader, whom she defines as "someone who cares enough about an issue to give some expertise in that matter and then share it with others." She thoughtfully remarks that she has "become that teacher-leader." Yet, the realities of accepting this role have brought about many unforeseen challenges, and Alice's leadership role has not been easily acquired:

> The offers to do more leadership activities came in my second year of teaching, but on the other hand, I'm not treated as a teacher-leader in my district, not in my site. When I was in one position, they didn't think of me as a teacher-leader. They certainly treated me as a rookie.

Rachel's zest and passion for students have moved her into a number of leadership roles in her school and district. Taking leadership has meant being uncomfortable by not necessarily agreeing with the status quo. She believes that taking leadership does not require being an "expert." Rather, it requires being able to recognize something as important and to be willing to work toward improving the situation. It is this belief that keeps her willing to stand up and do the scary stuff. However, this has not been easy for Rachel, who has seriously considered removing herself from the role. But then she wonders, if she quits, who will take her place? Being a leader is not something she had intentionally sought to do:

> I never wanted to be a leader, really. I like to be the person that does good things and fades into the woodwork. Because with leadership comes a certain amount of visibility and with that visibility come other things, too. Things that I necessarily don't want. But it boils down to putting your money where your mouth is. If you believe it, you do it. If you are going to wait for someone else to do it, you could wait a long time. Everyone wants to get along and be nice. I understand that, but I think I have to make a stand when it comes to these kids and be willing to stand up and be counted or have stones thrown at me or whatever. It's real important to do that because that's the only way to contest and contradict the way the system has been functioning and to make a change. I think it has to be made public because a lot of people don't even realize that there is a problem. [In the project] we say a leader is someone who takes responsibility of what matters to her. That has done more for me, that definition, than I can ever say.

GENDER, CLASS, AND RACE BIAS

Issues of race and socioeconomic class appear to outweigh Alice's and Rachel's concerns about gender issues. In reality, it is the complex mingling of these issues that makes it inappropriate to try to isolate gender. Examining and discussing gender alone will not help us understand how bias has affected women and girls of color.

Interestingly, important incidents in the lives of Rachel and Alice that have had strong implications for their equity work involve what happened to their brothers in school. Alice shared her emotions:

> I have examples of friends I knew who were so smart. That is another thing that rages inside of me. I talk about it all the time. I had a brother who was way better in math than me. Somehow he was rejected by the system. He didn't just drop out, he was forced out. This makes me more passionate and probably aggressive about telling kids the things that I wasn't told. Sometimes I think I get too passionate, and I get my feelings hurt when I don't get the response I want. I hurt a lot about that. I'm burning with this thing that I care about. If I could wake up one morning and not care about it, my life might be easier. I wouldn't have to do any of the things I do other than teach, go home, and get my check on the first. But this is the thing that I care about. It causes me to cry at night and it causes me to dance on the tables, too.

Rachel described her brother's experience:

> Going through school, I saw things that I knew weren't right. Different people were treated [badly], just because of their ethnicity or their handicap. That sensitized me quite a bit. I have this older brother. I consider him brilliant, incredibly intelligent, but he was never [encouraged] in school and he became very, very frustrated. He has read more than most college peers that I know. He can discuss incredible things. I see how he was hurt by the system that didn't recognize his intelligence and didn't reinforce it, didn't validate him and saw him as a loser. I see how that has happened to other people. That has always been an issue with me. I want to do something so that won't happen to somebody else.

As Alice continues to gain confidence in her expertise and knowledge, she seeks ways of working with others to bring about the changes that she believes need to happen. Unfortunately, Alice also continues to be saddened and increasingly embittered by the lack of support at her school site. In addition to attributing much of this negativity to her new-teacher status, Alice believes that racism is at work:

> I think my few years of teaching have something to do with it. I think that color has a lot to do with it. If we look at people who know mathematics, you wouldn't look at an African American, because we tend to score low or lowest. I think there is some baggage that is carried into the leadership world. So here is a person who traditionally hasn't scored high, wasn't in the AP classes, wasn't in the A-F classes, and here she emerges. [I think they wonder] "I never saw you in any of my classes." I think it has some backlash to it. So people don't see me as a mathematician. My teacher experience and my color stand in my way.

Racism affects individual as well as institutional responses to people of color. Selective perception and reinforcement are used to deny variability among people of color in such areas as intellect and accomplishment. Rachel believes that teachers' assumptions of children's capabilities, even when they come from a desire to help the child, can at times be misguided. Rachel talks about how racism is a topic around which so many educators are "incredibly touchy":

> People who have ideas that actually can be termed *racist* don't see it in themselves and they don't see it in the thought, the racist thought. It is like we are talking in two different dimensions here. They don't even understand what they are saying. If you ever told them that what they are saying is unfair or untrue or racist or whatever, they will be the first to deny it. It's there and I've seen it.

Alice and Rachel believe that the challenge is to provide rich mathematical experiences for all children at all levels and to not use the different challenges as excuses for not doing so as Rachel expressed:

> I think that the challenge is going to be for teachers who care to get the things together that can provide kids with the types of mathematical experiences that will really teach them math, and not some rote something that they'll never use again in their life. And to not water it down so much and to not assume because it's going to this kind of kid or with this type of language that it has to be that much different. It really doesn't have to be. They're totally as capable on all levels.

The issues of race and gender are foremost on the mind of Rachel as she ponders their influence in mathematics education:

> Why are most higher-level mathematicians basically of one gender and race? The more I think about the world and world issues and who has say, [the more I realize] it's going to be the people who have the skills and knowledge. It's given me another piece of the big picture. It's made me want to put more effort into my mathematics curriculum. I thought math was important but reading and language arts were more important. Now I see the role mathematics plays in creativity and enlarging the mind. I see mathematics as an important, vital tool for the liberation of any young mind. I think kids are frustrated and angry because they do not see their options and they have been educated with blinders on and they are not even aware of the potential that is out there. We need to teach them to be questioning thinkers. It's critical for their life, for their life survival, the survival of their psyche, their self-esteem, and everything else.

The Projects

Rachel and Alice have been participants and leadership-team members for projects that focus deeply on issues of equity and leadership in mathematics education: a statewide project funded by the California Mathematics Project and two other projects—a local systemic initiative and a national institute—funded by the National Science Foundation. In each project, at least 50 percent of participants are educators of color.

The projects' primary goal is to promote equity within mathematics education. The projects are based on (1) the mathematics-reform and educational-change literature, (2) the research on achievement of underrepresented groups, (3) experience in several earlier projects, and (4) more than a decade of consultation with teachers and educational leaders at state and national levels. The projects strive to create an infrastructure to support educators' addressing equity in mathematics education in ways that produce significant changes in the mathematical experiences of students from underrepresented groups (Weissglass 1996).

The Support Structures

Structures that provide emotional support to enable teachers to move along the stages of developing leadership form an essential component of the work done in the projects. In these structures, two or more individuals equally share listening and speaking time, without giving opinions or advice. These structures are rooted in the theory of constructivist listening, which views intelligence as a flexible and adaptive human characteristic. This theory holds that an individual is capable of solving his or her own difficulties if given enough attention as he or she talks about it. The costructivist listening structures fall into three categories:

- Dyads—the exchange between two people of listening and talking for a fixed period of time
- Support groups—the exchange of listening among a small group of individuals. Each person talks for a fixed amount of time while the others listen attentively
- Personal-experience panels—the opportunity for a small number of people to have a fixed amount of time to talk to a larger group about their personal experiences related to a particular topic

The constructivist listener aims to enable the talker to express his or her feelings, to construct personal understanding, and to use his or her full intelligence to respond creatively to situations rather than to rely on habits or old coping strategies (Weissglass 1990).

Alice has found that the support structures afford her an opportunity to be listened to and have shortened the "muddle period" she needs for solving a problem. Through her roles as a participant and a support-group leader, she has had an opportunity to think about how her feelings and emotions have changed as she takes on more and more leadership. She has found what she refers to as a "new fire for clarity" as she thinks about the ideas that have significance for "just about every issue" in her life:

> I know that one of the most important things that I have gotten is to be able to [be listened to as I] talk. I used to just like talking to myself by writing to myself. I was a journal writer, but now it is just like, "What do you think of this." So I learned how to get support for myself and to reach out to people. If I am talking to you, it is real interest that I have. I'm really trying to get something for myself and I hope that I give something back.

The emotional support that Alice and Rachel have received in the projects has affected their ability to bring about changes in their classrooms. Alice has taken this support back to her students. She now spends more time encouraging them to listen to, and talk with, one another. Although she hasn't quite found a way to replicate support groups, she regularly uses dyads with her students:

> I do tell my kids, "You're going to make it based on the support you make right now, the alliances you make right now. I know you and we're together and we can pull from each other." So I try, with every cooperative group, to get some alliances built so that they are really dependent on each other, supporting each other. But I am just beginning.

Rachel appreciates the support structures and the opportunity to discuss and share important issues with colleagues:

> I feel stronger in terms of my beliefs and what I'm doing and in my ability to discuss these issues. I feel that my opinions are not way out. They're actually valid opinions, and I have a right to share and express them and [provide] some insight to people. So it's been really validating to me.

The support structures are based on the belief that distress, or the accumulation of unreleased emotions left from hurtful experiences, is a primary source of unintelligent and uncaring behaviors. The natural physiological process of emotional release contributes to the recovery from the effects of distress, which leads to clearer thinking. Alice has experienced the benefits of emotional release around the issues that affect her leadership. From the moment she was introduced to this concept, she felt it ring true for her. As she found ways to release her emotions, she understood more clearly the leadership role she would choose to take. Alice believes that it has enabled her to find the truth that was already inside of her:

> I believe that by working with teachers, I can change them and then have an impact on more kids. I don't want to lead just to lead. I like to lead because I want the desired effect. I want my brother and my sister to be in the classroom and to feel comfortable there. That's my wish and desire.

Rachel describes the support she has received as necessary for her to continue in her work. She believes that people who are attempting to change schools must have support. She realizes that it is very easy for her to get caught up in the day-to-day challenges and site and district issues that affect her work. The support structures help her clarify her thinking and facilitate her ability to make decisions in a rational way:

> The structures foster a lot of safety in sharing. That is to me one of the best things about [the project]. People are able to reveal themselves in a very deep level yet feel safe. Confidentiality and the rules of taking turns when we speak [foster] mutual respect. People are seen as equals, with nobody having any more right to speak than another person. The rules have built in the underlying message of equity and respect for people.

An Issue of Trust

Trust, another component addressed in the projects, has particular significance for Alice and Rachel. It has helped them take action. Before coming to the projects, Alice was not aware of the important role that trust plays in relationships. She now views trust as the primary component of the work she does and uses questions about trust as her guide for deciding the work she is willing to take on. In the projects, she has experienced the power of talking about a problem with several people she trusts, and she feels empowered to act on her decisions. Through discussions with trusted individuals, she can talk about her actions and explore new ideas. This situation builds the supportive environment that she needs to contradict the unhappy situation she finds at her school site:

The word came back today over and over again, *trust*. [I need to] stop putting my confidence in nontrustful situations, then I won't have to spend a lot of time healing myself. . . . This situation that I am into now I got into before I met [the projects] and if I would have asked these questions [about trust], I would have never got into it. All the signs were there that they are not trustful folks. I could breathe fire out of my mouth and nose because the signs were there and I shouldn't have trusted, but I did. And so now everything is about betrayal, as sure as the night follows the day.

Alice believes that her assumptions and beliefs about learning, teaching, and the purpose of school and about issues of race, class, and gender bias have been "substantiated." The discussion in the projects confirmed what she already "knew to be true." Rachel wonders about the superficiality of the discussions among teachers at her school. They don't engage in the conversations that would facilitate their sharing assumptions and beliefs about teaching and learning:

That's not the type of thing you talk about. So it's hard to connect with people who have maybe similar goals, similar ideas and philosophies, unless you happen to hit it off at some other level and you become friends and you talk. And that can happen, because that has happened, and I have some good friends that have happened just that way. You can feel real isolated if you're not around other people who feel that way or are willing to be vocal about their feelings. You can feel real isolated. It has been wonderful for me to be a part of [the project]—to have these issues shared and discussed openly, in an intelligent manner, with respect for people's diverse opinions.

SUMMARY

Women of color frequently enter educational leadership through special-project positions and are, in general, assigned to work with minorities or on minority issues. Although this work is good and necessary, it does not prepare women of color to work with white children and adults. In similar fashion, most white women are not given an opportunity to develop the skills and knowledge necessary to work effectively in addressing issues with people of color. Because of this systemic isolation of people of color from whites, it is necessary to institute structures that facilitate open and honest dialogue between the groups. It requires a safe environment in which educators can reveal, discuss, and change their assumptions about people from underrepresented groups.

Alice continues to be a potentially strong force for bringing about change in schools. Her involvement in the projects gives her the emotional support she needs through the structures and the trusting relationships she has acquired with people of backgrounds different from hers. She also values the knowledge that she has gained about constructivist and social-constructivist mathematics while focusing on issues of bias. Alice has decided to avoid working in distrustful environments and actively seeks to work where her knowledge and expertise are welcomed and supported. She continues to rely on emotional support during the difficult times as she works to improve schools for students very much like herself.

Rachel finds herself more willing to speak up to her colleagues and more willing to approach anybody—an administrator, an aide, a parent, or whomever—when she believes that a situation calls for an issue to be discussed or raised: "I don't have as much hesitancy at all about bringing up a lot of issues and raising them in regard to equity or appropriateness of the curriculum or something like that."

To bring about a better understanding of how issues of race and socioeconomic class affect the lives of women, structured opportunities for honest discussion in safe and trusting environments will be necessary (Weissglass 1991). As educators address issues of educational change, particularly in regard to equity, they will need emotional support in addition to the content and pedagogy for meeting the challenges they will encounter. Emerging teacher-leaders need support to prevent the discouragement that generally leads to burnout. Finally, for the mathematics reform community to address adequately the achievement of students of color, we must hear the voices of many experiences in the conversation. Emotional support gives strength and clarity to these voices.

REFERENCES

Banks, Cherry A. M. "Gender and Race as Factors in Educational Leadership and Administration." In *Handbook of Research on Multicultural Education,* edited by James Banks and Cherry A. M. Banks, pp. 65–80. New York: Macmillan, 1995.

Edson, Sakre K. "Voices from the Present: Tracking the Female Administrative Aspirant." *Journal of Educational Equity and Leadership* 7, no. 4, (1987): 261–77.

McCarthy, Cameron, and Warren Crichlow, eds. *Race, Identity and Representation in Education.* New York: Routledge, 1993.

Weissglass, Julian. "Constructivist Listening for Empowerment and Change." *Educational Forum* 54 (Summer 1990): 351–70.

———. "Teachers Have Feelings: What Can We Do about It?" *Journal of Staff Development* 12 (Winter 1991): 28–33.

———. *Ripples of Hope: Building Relationships for Educational Change.* Santa Barbara, Calif.: CECIMS, University of California, 1996.

A Latina Tale

The Experience of One Latina Mathematician

11

Linda Valdés

When a colleague in the mathematics and computer science department asked me whether I would be interested in writing an article about Latinas and mathematics, I was at first perplexed. I am a Latina university professor, but what, outside of mathematics, could I tell an audience of teachers that they would find of interest or want to know? I did not feel qualified to tell teachers how or what to teach, so what could I contribute? After some thought, I decided to share my educational experiences in the public school system. Even so, a question remains about the relevancy of the experiences of a woman who graduated from high school in the 1960s. Teachers are now more culturally sensitive and aware of the differences in their students. But perhaps the experiences of a Latina from a traditional Cuban American household have not changed and might be enlightening to those who seek to know more about the Hispanic culture. I use the words *Latinos*, *Hispanics*, and *Spanish* interchangeably. At the risk of offending some, I have yet to determine the differences among the words and become more confused on the concepts when someone endeavors to explain them. In my own defense, my family and Hispanic friends are also mystified.

If there is one aspect of the Latino culture that best defines it, it is the family. We need to understand that there are various levels of the Latino family: the parent-children unit; the extended family that includes *abuelos*, *tíos*, and cousins; and the neighborhood, which in most cases is predominantly Hispanic. In my experience, the traditional Hispanic family is one in which the father works and the mother stays at home and cares for the children. The family revolves around the children, who are the chief concern of the parents. The father works forty to sixty hours a week, and the job of rearing and disciplining the children falls mainly on the mother. The father retains veto power on any decisions, but it would be unwise for the commander-in-chief to disagree with his general in charge of the troops. The mother is responsible for keeping in touch with the relatives. Birthdays, anniversaries, births, and holidays are celebrated at the homes of different relatives. It is during these occasions that family matters are discussed and opinions are shared.

A girl who grows up in a Hispanic neighborhood feels a great sense of comfort and safety. Here are people like herself, who speak the same language, eat the same foods, and celebrate the same holidays. The people in the neighborhood may have similar names: Jiménez, González, Domingues, Andrade, Fernández. The children usually go to the same school. The mothers talk and work together and are there for one another in times of need.

At some time during the young life of a Latina, she notices that there are "others" living outside the neighborhood, with different-sounding names. She also realizes at some point that the fathers seem to have different jobs, drive different cars. She realizes that her doctor, dentist, and most of her teachers are not from neighborhoods such as her own and do not have Spanish names. There is no resentment, just an awareness. At the dinner table, we used the word

Americans to distinguish these people from ourselves. My parents, aunts, and uncles were born in the United States, yet they called themselves the Spanish; the Anglos or whites were the Americans. Were we tacitly acknowledging that some groups in this country had more advantages and access to opportunities than others? I do not know, for we never discussed it, but I believe this was so.

The elementary school that I attended had children from the surrounding working-class neighborhoods with white and Hispanic families. My best friend was white, and our mothers became friendly and often shared the job of transporting children from here to there. I remember my first encounter with mathematics in the first grade. The teacher was asking out loud, "What is 1 plus 2? What is 2 plus 3?" I did not have a clue what she was talking about, but other students were raising their hands, answering the questions, receiving praise, and enjoying themselves. I decided that I wanted the same for myself. The Latino child during these early school years is easy to reach. If an elementary school teacher wants to encourage such a child, the most effective method is commenting, "Your family can be very proud of you." The most damaging comment would be "You are shaming your family." A teacher will find the child's family eager to work with him or her. The parents will want to be informed about how to help their child learn to live a happy, full life. Their primary goals are not to make their child overly competitive or wealthy or ambitious. They would be especially alarmed if they thought that a teacher was encouraging a daughter with these ideas.

During these formative years, a Latina becomes aware that she is treated differently from her brothers. Young males are allowed freedoms that she will never be granted. It is not until she reaches puberty that she protests. "Why can my brother stay out later than I can? Why doesn't he have to wash dishes, make the beds?" The answer she will receive will invariably be "But, your brother is a boy, and that's the way things are. It's not fair, but it's not unfair. It's just the way life is."

By junior high school, she knows that she should never push herself forward, should never put herself in a position in which she would have to address a group of people or accept any kind of position of authority other than that of a nurturing, supportive kind. It is permissible, however, for her to speak from her seat. At family gatherings, there is much talking and different levels of conversation. The level at which all members of the family listen to the speaker is male dominated. But females are allowed their input, usually in the form of teasing or scolding. When the Latina does speak, the man who thinks that he posseses her will roll his eyes or lift his arms to heaven as if imploring, "Why me, God?" Perhaps more of import is communicated by the women in these short outbursts than at any other time during the gatherings.

In class, when Latinas do speak up, they have something to say, and it is very thoughtful. I remember scolding my seventh-grade geography teacher one day for favoring the boys over the girls. I learned that he was somewhat taken aback by this behavior because four years later, when my sister was in his class, he asked, "Are you anything like your sister? Are you going to try to lead the girls out of Egypt, too?" My sister was quite embarrassed and let me know as much. I, however, was surprised and somewhat gratified. He had been listening— my words had not been dismissed.

A teacher can expect the full support of Hispanic parents. I learned that early on. I knew better than to come home with complaints about my teachers. The teacher's behavior was always appropriate and mine was always suspect, so it was to my advantage not to complain. It follows that the child listens to her teacher and believes what she is told. Teachers need to consider this a heavy responsibility. Remarks made by teachers to their young charges are heard, remembered, and referred to over and over again. If a child is energetic and inquisitive and the teacher rebukes her for that curiosity, there may be unhappy consequences.

It is not expected that the Latina should excel. This sentiment is shared by most, including her classmates. I attended junior high school with a nice Anglo student named Jesse who was outstanding in everything. He was a star athlete, he made good grades, and he was popular. But I made better grades. This was disconcerting to a young man who was accustomed to being the best of everything and considered it his God-given right to be so. One day he approached me and said, "Sure you're smart in school, Linda, but you have no common sense." I worried a great deal about that statement because I believed this young man. How was it that I had no common sense? Why would he tell me this if it were not so? But it was rather strange. We did not share many conversations. Why should he feel it necessary to have one in which he was hurtful? This kind of behavior toward minorities and women still occurs. The system is set up so that we cannot compete fairly. When those who have been preordained to succeed discover that *others* are competing with them successfully, they see an alarming problem. This young man and others like him feel compelled to find fault with minorities and women—and they must share that knowledge with us. Otherwise, the natural order of things would be disrupted, and the unwritten promises of privilege would not be kept.

By the time I arrived in high school, all my close friends were Latinas. Those friends with whom I spent time before school, after school, and during lunch were Latinas. Those friends with whom I spent hours gabbing over the phone were Latinas. I am not sure how or when the transition occurred, but some time between elementary and high school, socialization and segregation occurred. A teacher who uses cooperative and group learning in his or her classes might want to reflect on this situation. If a Hispanic, male or female, is in a class with mostly white students, he or she is at a disadvantage in the group situation. Such a setting probably is not as comfortable for the Hispanic student as for the white student. There will, in general, not be an easiness in trading phone numbers because often the Hispanic student does not want to invest energy in working with others when there is a tacit understanding that she or he will always be at some disadvantage. I am not trying to imply that there is some kind of vindictive spirit or conspiracy in which those who are in the privileged categories are out to get those who are not. For the most part, I did not find this to be true. We do need to acknowledge, however, that there is such a thing as a "comfort zone" within which students are more inclined to work cooperatively because they do not feel like outsiders.

Why do we naturally segregate ourselves? There are many reasons. Comfort is one. It is obvious that if we are made to feel awkward or ignorant in some group's presence, we will try to avoid that group. For instance, Hispanics are not well versed in sarcasm. I was often at a loss to understand what Anglos were telling me when they said something sarcastic. I could not comprehend the meanings of a sentence when the words meant one thing, but the sentence meant something quite different. I have difficulty with this even today. We should not expect humor from one culture to be understood by another. Even though it is just humor and all in good fun, certain humor may be viewed as insensitive, hurtful, and even insulting by those from another culture. I remember how I used to sit in class and listen to jokes and sarcasm, hear my classmates and teachers laugh, and wonder what was so funny.

I went to junior and senior school high during the 1960s, the era of the fight for civil rights that was led by African Americans. The schools in my city were segregated: white, Spanish, and black. Although there was some mixture of whites and Spanish in the high schools, there was never a question of the cultural identity of the school's population. With the movement for school integration, the parents in the Latino neighborhoods were given the opportunity to choose where to send their children to school. I could attend the high school close to where we lived, and from which my mother graduated, or a white high school. My parents sent me to the white high school. Does the school to which students go matter? Of course it does. It was not a surprise when a student from my alma mater continued on to college; it was the exception for the graduate of my mother's alma mater. The school matters, particularly for Latinas.

As with many Latinos, no one in my entire extended family had a university degree. I had no base of experiences from which to draw. There was little reason to expect that I would choose to go to college. It takes some courage to display the kind of pioneer spirit needed. When a Latina sees her peers preparing for college and knows, because her grades tell her so, that she too is college material, she is helped in her resolve to further her education. If she attends a predominantly Hispanic high school, where few students (mostly males) go to college, it is all too likely that we will lose her talent.

One day at supper I announced to my parents, "I want to go to college." This was pretty momentous for my family and myself. None of us knew what this would involve, but we did know that it was an opportunity. My parents could not call on the larger family for information because the facts were not there, either. But advice certainly was. I remember my parents explaining apologetically to my relatives why they decided to send me to college: "We want her to have an education to fall back on just in case she marries a bad man, and she has to support her children." This statement describes very well the perception that Latinos have about their daughter's future: (1) It is very likely that she is not wise enough to choose a good man to marry. (2) She will have children who will be central to her life. (3) A career is of little importance. Nevertheless, my parents were determined to do the best they could for their eldest daughter and scrimped, saved, and took out loans to send me to college. They were always ready to make sacrifices for their children. Even now, in their retirement, they fret and worry how they can best leave their savings to their children after they have passed away.

My high school offered a special class in mathematics. If students scored above a certain level on a test, they could participate in an accelerated program and would continue with the same teacher and students from the tenth through twelveth grades. There were three Latinos in class, two boys and I. I was content with the class; I enjoyed the subject. But I never listened to any lecture given by Mrs. T., my mathematics teacher. When she began a lesson, I would open my book and read the section myself and do the problems. I am not sure why I believed that I was better off doing mathematics on my own. I felt no dislike for my teacher, no hostility, but there seemed to be little communication between us. There was also little encouragement and little attention. I am not trying to cast any blame on the teacher, and at the time I probably was relieved that she was giving me little thought.

Every year, the Florida Senior Placement Test is given throughout the state to all seniors. We received our scores one morning during homeroom. During mathematics class, Mrs T. asked how many of us had placed in the 99th percentile. Five of us raised our hands. "You did, Linda?" Mrs. T. asked in a shocked voice. I was surprised at her tone and thought that perhaps I had not been reading the scores correctly, so I looked again and replied that I had. A few seconds of silence passed. "Will everyone please pass forward your scores." I will not try to guess what was going on in her head. Maybe she refused to believe me and had to read the reports herself. Nor can I hazard a guess about what she thought when she realized that this little Spanish girl, whom she had barely noticed over three years, could be excelling in the subject she was teaching. I received an A on my report card for the last grading period. I was careful in calculating my grades, and I knew that I should have received a B. I do not reflect back with any bitterness toward that teacher, but I do know that I learned mathematics not because of her, but in spite of her. And that is sad.

Empowering Young Women in Mathematics through Mentoring

12

Dorothy Buerk
Ann Oaks

And Kelly was beginning to see that Sarah didn't let anything stop her, she just went and did it. This whole attitude of not letting other people stop you, of believing that you could do what you wanted, was something she found herself imitating.

—Michael Crichton, *The Lost World*

Kelly Curtis, a fictional thirteen-year-old girl, chose a successful woman scientist, Sarah Harding, as her idol and role model. In the course of their interaction, Sarah became Kelly's mentor. Their story is described by Michael Crichton in his novel *The Lost World* (1995), a sequel to *Jurassic Park* (1990). In this chapter we will describe the mentoring relationship between Sarah and Kelly and discuss how you can mentor the young women in your care.

Our young women need mentors and role models to help them see their potential, raise their aspirations, and work for their goals. They need mentors who reach out to them and encourage, support, and guide them. Young women need to see female role models in the careers to which they aspire; they may also look for them in books as well as in their personal worlds. You also may find ideas about mentoring in the literature you read. Sometimes you will find those ideas in very surprising places, as we did when we read *The Lost World*.

ROLE MODELS AND MENTORS

As a young woman progresses through her education, we hope that she will have both role models and mentors to guide and influence her. Role models are people who show young women possibilities and exemplary behavior and are people they can emulate and imitate. Sometimes a young woman chooses her own role models, often without their knowledge. We try to provide role models by inviting women who use mathematics and science in their professions to meet the young women in our care.

Mentors, however, need to have personal interactions with the young women. Mentors are people—

- who help us expand our vision of what is possible for us, especially as we venture into realms that are beyond our experience or different from society's gender-role expectations of us;

- who may have more faith in us than we have in ourselves and encourage, even push, us;

- who have experienced what we are now experiencing, or hope to experience, and so can guide us, warn us of obstacles, and help us stay on course when we are tempted to veer off;

- whom we respect and want to emulate and imitate.

THE CHALLENGE THAT YOUNG WOMEN FACE

Young women receive many messages that devalue them. We need to be vigilant in countering those messages in our interactions with them. Peggy Orenstein (1994) describes the experiences of middle school girls and their struggles to avoid silence; to become confident; and to gain the respect of their peers, parents, and teachers. Their quest to gain the respect of others is often in conflict with gaining their own self-respect. Orenstein, speaking to women, presents the problem that young women face as she describes the culture in which we live and in which these young women are growing up.

> Girls with healthy self-esteem have an appropriate sense of their potential, their competence, and their innate value as individuals. They feel a sense of entitlement: license to take up space in the world, a right to be heard and to express the full spectrum of human emotions. The fact that, in study after study, women and girls are less likely to feel those things than men and boys should be no surprise. We live in a culture that is ambivalent toward female achievement, proficiency, independence, and right to a full and equal life. Our culture devalues both women and the qualities which it projects onto us, such as nurturance, cooperation, and intuition. It has taught us to undervalue ourselves. Too often we deride our own abilities. We denigrate our work and discount success. We don't feel we have the right to our dreams, or, if we achieve them, we feel undeserving. Small failures may confirm our own sense of inevitable failure, making us unable to take necessary risks. We learn to look outward for markers of acceptability. (Orenstein 1994, pp. xix–xx)

Orenstein has presented the problem that we face as teachers working with young women. Although many of our young women have the confidence to manage well as they maneuver through the negative messages that they hear about their worth and their potential, many others are less fortunate. Once we become aware that young women tend to look for external signs that what they are doing is acceptable, we understand the extreme importance of making mentors and role models available to them.

Mathematics is often considered such a male domain that many young women are discouraged from excelling in mathematics even when they have the desire, the ability, and the background. Since mathematics is seen as a male-oriented discipline, we often find it easier to let our young women give up when they have difficulty in the subject rather than persist in it. Research has long held that it is more difficult for females to seek and value success in a domain often considered male. In one classic study, Horner (1971) found that the perceived possibility of negative consequences from success in a male-associated field inhibited some women from achieving in that field.

Our young women, and indeed our young men as well, need to know that women are, and continue to be, successful in mathematics and in the sciences. Our young women need to know that they can and should pursue their aspirations for success in these fields. Both women and men can help these young women reach their potential in our culture, which is still struggling to attain real gender equity.

KELLY AND SARAH: A FICTIONAL TALE

With that background, let us share the story of the interaction between thirteen-year-old Kelly Curtis and Sarah Harding, her mentor. Kelly was poor (her clothes were hand-me-downs from her sister) and she was bright. She had only one real friend in school; he was "the only person who thought it was okay that she was smart" (Crichton 1995, p. 46).

Kelly had a heroine, Sarah Harding, a young woman scientist who had been a poor scholarship student at the University of Chicago and was now (in Kelly's words) "the most famous young animal behaviorist in the world" (Crichton 1995, p. 61). Kelly had read every article she could find about Sarah Harding's life. She had studied pictures of Sarah very carefully. Kelly found Sarah "beautiful and independent, a rebel who went her own way" (Crichton 1995, pp. 61–62). Kelly found a person with a similar background who had achieved success in science. She had chosen a role model and studied her carefully.

In the course of the novel Kelly and Sarah meet. Although this was exciting for Kelly, she worried about the impression she might be making on Sarah. At one point Sarah and Kelly had this conversation (Crichton 1995, p. 231):

"What do you like to study?"

"Actually, uh, I like math," she said, in a sort of guilty voice.

Sarah must have heard her tone, because she said, "What's wrong with math?"

"Well, girls aren't good at it. I mean, you know."

"No, I don't know." Sarah's voice was flat.

Kelly felt panic. She had been experiencing this warm feeling with Sarah Harding, but now she sensed it was dissolving away, as if she had given a wrong answer to a disapproving teacher. She decided not to say anything else. She waited in silence.

Kelly had learned to be silent when she feared that she was wrong or had displeased. Silence is a strategy that many of our female students have learned to use when they feel that their words do not initially please the listener or that their words are not being heard. This silent behavior sometimes becomes a strategy that is hard to break.

After a moment Sarah ... sat down and started putting on a pair of boots. She moved in a very normal, matter-of-fact way. "What did you mean, girls aren't good at mathematics?"

"Well, that's what everybody says."

"Everybody like who?"

"My teachers."

Sarah sighed....

"And the other kids call me a brainer. Stuff like that. You know." Kelly just blurted it out. She couldn't believe that she was saying all this to Sarah Harding, whom she hardly knew at all except from articles and pictures, but here she was, telling her all this personal stuff. All these things that upset her.

Sarah just smiled cheerfully. "Well, if they say that, you must be pretty good at math, huh?"

"I guess."

She smiled. "That's wonderful, Kelly." (Crichton 1995, pp. 231–32)

At last Kelly hears an adult whom she respects reassure her that it is OK to like mathematics. But Kelly had heard other negative messages as well.

"But the thing is, boys don't like girls who are too smart."

Sarah's eyebrows went up. "Is that so?"

"Well, that's what everybody says..."

"Like who?"

"Like my mom."

"Uh-huh. And she probably knows what she's talking about."

"I don't know," Kelly admitted....

"So she could be wrong?" Sarah asked, glancing up at Kelly as she tied her laces.

"I guess."

"Well, in my experience, some men like smart women, and some don't. It's like everything else in the world." (Crichton 1995, p. 232)

Sarah challenges Kelly to question what she hears. Could Kelly's mother be wrong about whom boys will date? Might Sarah be the authority on smart women and their relationships with men? Sarah continues the conversation by talking about situations in which people have been misinformed (Crichton 1995, p. 233):

"So, Kelly, even at your young age, there's something you might as well learn now. All your life people will tell you things. And most of the time, probably ninety-five percent of the time, what they'll tell you will be wrong."

Kelly said nothing. She felt oddly disheartened to hear this.

"It's a fact of life," Sarah said. "Human beings are just stuffed full of misinformation. So it's hard to know who to believe. I know how you feel."

"You do?"

"Sure. My mom used to tell me I'd never amount to anything." She smiled. "So did some of my professors."

"Really?" It didn't seem possible.

"Oh, yes," Sarah said.

Sarah had a mom who did not see her potential. Kelly did as well. Sarah had some professors who did not think that she would amount to anything. Kelly had some teachers who discouraged her interest in mathematics. Sarah was like Kelly in more ways than Kelly knew. Sarah had been discouraged also, but that had not stopped her. Kelly had met the heroine about whom she had read. Sarah was real, very matter-of-fact. Sarah seemed to understand Kelly, who found herself blurting out things that she did not intend to say. Sarah told her that many of the messages she had heard were wrong. Kelly was not sure she wanted to hear that. But Kelly had studied Sarah Harding's life because Sarah had been a smart girl from a poor family who became a successful scientist. Sarah was a role model for Kelly; Sarah was becoming a mentor for Kelly.

Kelly got to know Sarah better through their interactions and adventures, and she came to respect her even more. Kelly realized that Sarah followed through on her ideas in spite of what anyone else said. Kelly realized that "this whole attitude of not letting other people stop you, of believing that you could do what you wanted, was something she found herself imitating" (Crichton 1995, p. 341). Kelly also remembered that she should not believe everything that she heard and began to question things she heard, even things she heard from Sarah Harding. At one point, when there seemed to be no solution to a problem, Sarah said,

"Okay. So we have two hours to get to the [helicopter] pad."

Kelly said, "How can we do that? The car's out of gas."

"Don't worry," Sarah said. "We'll figure something out. It's going to be fine."

"You always say that," Kelly said.

"Because it's always true," Sarah said. (Crichton 1995, p. 346).

As you may suspect, from your knowledge of either the book or the movie *Jurassic Park*, this novel is a series of adventures and life-threatening situations that are far from real. These situations, however, accentuate the interactions between these two female characters. We must add that you will find none of this interaction in the 1997 movie, *The Lost World: Jurassic Park*.

Sometime later, part of the group found themselves in a tight situation from which they needed to escape. Kelly knew that the computer contained maps of the area, and the others present encouraged her to try to find them. Kelly could not find the maps.

On the screen, she now saw a rotating cube, turning in space. Kelly didn't know how to stop it…. Kelly stared at the cube on the screen feeling hopeless and lost. She didn't know what she was doing anymore…. But she couldn't focus. She couldn't click on the icons, they were rotating too fast on the screen. There must be parallel processors to handle all the graphics. She just stared at it. She found herself thinking of all sorts of things—thoughts that just came unbidden into her mind.

The cord under the desk.

Hard-wired.

Lots of graphics.

Sarah talking to her in the trailer….

In the trailer, Sarah said: Most of what people tell you will be wrong….

She kept thinking of the cord under the desk. The cord under the desk. Her legs had kicked the cord under the desk.

[Someone] said, "It's important."

And then it hit her.

"No," she said, "It's not important." And she dropped off the seat, crawling down under the desk to look.

"What are you doing?" [someone] screamed.

But already Kelly had her answer. She saw the cable from the computer going down into the floor, through a neat hole. She saw a seam in the wood. Her fingers scrabbled at the floor, pulling at it. And suddenly the panel came away in her hands. She looked down. Darkness.

Yes.

There was a crawlspace. No, more. A tunnel.

She shouted, "Here!" (Crichton 1995, pp. 383–84).

The answer was not in the computer; it was in the way the computer was hooked up. She found a tunnel. Kelly had found a way out, but not by looking in the place where everyone believed she should be looking.

Sarah Harding, who had not been with this group, met them soon after their escape through the tunnel that Kelly had found. The following conversation ensued.

> "So it was the graphics that gave you the clue, Kelly?" Harding said, admiringly.
>
> Kelly nodded. "I just suddenly realized, it didn't matter what was actually on the screen. What mattered was there was a lot of data being manipulated, millions of pixels spinning there, and that meant there had to be a cable. And if there was a cable, there must be a space for it. And enough space that workmen could repair it, all of that."
>
> "So you looked under the desk."
>
> "Yes," she said.
>
> "That's very good," Harding said. "I think these people owe you their lives."
>
> "Not really," Kelly said, with a little shrug.
>
> Sarah shot her a look. "All your life, other people will try to take your accomplishments away from you. Don't you take it away from yourself." (Crichton 1995, p. 385)

Kelly had chosen Sarah Harding as a role model—a person who, like Kelly, grew up poor and smart and female. And Sarah was a successful scientist and a real person. Sarah was not aware that she was Kelly's role model. We often don't know when we have been chosen as role models. Through their adventures and interactions, a personal relationship developed between Sarah and Kelly, and Sarah became a mentor for Kelly. Mentoring is always a role of choice. Kelly now had a mentor who helped her gain confidence in herself, who helped her realize that it was all right that she liked mathematics and was good at it. She had a mentor who helped her realize that much of what she was told was not true. Kelly had a mentor who was a confident, creative problem solver and whose confidence in herself and in Kelly was clearly warranted. Kelly had a mentor who insisted that Kelly respect and accept her own accomplishments and not devalue them.

IMPLICATIONS FOR THE REAL WORLD

Unfortunately, our young women continually receive messages that devalue them. A fundamental quality of good educators is their respect for their students. We need to be sure that young women recognize that respect. They need to know that we expect them to respect and value themselves, their accomplishments, their talents, and their strengths and to realize that failures are important ways in which they can learn about themselves. Sarah would not let Kelly shrug off her success in finding the tunnel. She insisted that Kelly acknowledge her success and value it.

Hansen, Walker, and Flom (1995), in their report on what works for girls, address our role as mentors. They urge us to help girls and young women broaden their aspirations. The report says, "Agency, or the power to act, is important because it involves the process of translating one's dreams into reality—by striving, overcoming obstacles, and actualizing one's ideals" (p. 46). Many young women do not feel this agency, this sense of empowerment to pursue their aspirations. We can help by mentoring them. We need to help our young women believe that they *can* perform mathematical tasks, that they can realize their dreams. Kelly had her aspirations—to be a scientist like Sarah Harding. Kelly knew what she wanted, but she was getting too many messages that told her those aspirations were not appropriate for her. The discouraging messages and taunting by Kelly's peers sometimes shook her confidence. Even in 1995, she needed to hear that it is OK for a female to like mathematics. Kelly clearly needed support to keep believing that her aspirations were realizable for her. She got that support from Sarah Harding.

Many girls and young women need mentors, as Kelly did, because they are hearing much that is not true for them or their futures. Probably most damaging are the messages that stereotype women because of their gender, their race, or their ethnic background. Many of these negative messages try to hold the young woman back, asking her to settle for less than her dreams. In some extreme situations, these negative messages prevent her from even having any dreams. These messages might come from peers, from parents, from teachers, from counselors, or from other adults who—

- do not value academic success;
- are scared by creativity and unexpected ideas;
- think that they are protecting the student, often because she is female;
- have only a limited vision of what is possible, or at least of what is possible for females;
- are not really listening to the young woman's aspirations.

We need to be alert to young women getting these messages and present alternative messages to them.

STRATEGIES FOR
EMPOWERING
YOUNG WOMEN
IN MATHEMATICS

Let us suggest how you can become a mentor. The following strategies may prove helpful.

Establish the Relationship

- Seek out young women who need mentoring; help establish a sense of community among women.
- Be aware that some of the young women in your care have already chosen you as a role model. Be conscious of the messages you convey.

Implement the Relationship with a Focus on Yourself as the Mentor

- Remember that messages from peers and our culture are very powerful. Your words must be strong to counter such inaccurate messages as "Girls can't do math!" Do not be afraid to correct yourself immediately if you happen to convey a negative message.
- Point out inequities and messages that devalue women. Remember that your silence can be a very strong and negative message.
- An important message to convey is that you expect young women to do well.
- Share your own experiences in, and excitement for, mathematics.
- Be honest and forthright, but be as supportive as Sarah was with Kelly. Do not pretend that there will be no obstacles or that the road will always be smooth.
- Provide guidance. Because you have more experience, you can suggest next steps, show opportunities, and recommend special programs and educational pathways.

Implement the Relationship with a Focus on the Young Women You Mentor

- Let young women know that you expect them to be engaged, competent mathematicians. Encourage them to work hard, to experiment, and to explore alternative methods by providing an environment that supports exploration and discovery and encourages them to develop and express their own ideas. In a classroom situation, you can provide hands-on activities and group activities. Monitor these to see that all students are active in these settings. Be sure that females are not passively watching the males do the task.

- Do not let female students give up too easily. Do not be too helpful by giving answers prematurely. Sunny Hansen explains that when girls are not allowed to work their way through problems on their own, when they are simply given solutions, they are likely to develop a sense of helplessness—often called "learned helplessness" (Hansen, Walker, and Flom 1995).

- Listen for the seeds of ideas and encourage young women to pursue and develop them. Remember that without your support, their lack of confidence may cause these young women to ignore a good idea.

- Help young women raise and maintain their aspirations. Acknowledge their successes. Do not let them discount those successes, but be honest. Hollow praise is of no value and may, in fact, be damaging.

You will notice that many of these behaviors are consistent with the teaching strategies recommended by the reform movement in mathematics education and the NCTM's (1989) *Curriculum and Evaluation Standards for School Mathematics*. In fact, those of you who have been working on gender-equity issues in mathematics for many years will recognize them as strategies first developed as feminist pedagogy and used in coeducational classes in the early 1980s. We were pleased to see these strategies mainstreamed as valuable for all students with their inclusion in the *Standards* and the reform movement. Some of the strategies remind us that we must be vigilant to the messages that devalue women in our society and to speak out to counter these messages. They also remind us that our female students are more cautious in their interactions with us and that we need to help them feel free to share their ideas with us. We need to listen to those ideas with care.

As mentors, we need to be authentic, real, and honest. We need to share both the possibilities and the obstacles. Many young women try too hard to please, and many strive for a level of perfection that is not attainable. We need to model reality, not perfection. We need to talk about our own struggles and not minimize the difficulties we have experienced. The young women studied by Brown and Gilligan (1990) make clear their need for authentic relationships with the women who teach and advise them. They do this in a very powerful way.

Think about a young woman you have noticed, possibly in your classroom or another educational setting, or in your neighborhood, or in the family of a friend or a colleague. Perhaps she needs some encouragement or support to follow through on her ideas or to set higher academic or intellectual goals. Perhaps she needs encouragement to continue with mathematics. Perhaps she needs support because she is not confident in her mathematical ability. Could she be helped by your sharing of your own experience? Could you be a mentor for her?

Our young women have many opportunities in a culture that in some, but not all, ways has become more open to gender equity. We must continue to help our young women see opportunities, believe in their talents, and pursue those talents. We are often role models for students without our knowledge. We must be conscious of that responsibility. We become mentors only by choice. Sarah's mentoring of Kelly gives us a powerful model. We urge you to seek out young women to mentor.

REFERENCES

Brown, Lyn Mikel, and Carol Gilligan. *Meeting at the Crossroads: Women's Psychology and Girls' Development.* Cambridge, Mass.: Harvard University Press, 1990.

Crichton, Michael. *Jurassic Park.* New York: Alfred A. Knopf, 1990.

———. *The Lost World.* New York: Alfred A. Knopf, 1995.

Hansen, Sunny, Joyce Walker, and Barbara Flom. *Growing Smart: What's Working for Girls in School.* Washington, D.C.: American Association of University Women Educational Foundation, 1995.

Horner, Matina. "Femininity and Successful Achievement: A Basic Inconsistency." In *Roles Women Play: Readings toward Women's Liberation*, edited by Michelle Garskof, pp. 97–122. Belmont, Calif.: Brooks/Cole, 1971.

National Council of Teachers of Mathematics. *Curriculum and Evaluation Standards for School Mathematics.* Reston, Va.: National Council of Teachers of Mathematics, 1989.

Orenstein, Peggy. *SchoolGirls: Young Women, Self-Esteem, and the Confidence Gap.* New York: Doubleday, 1994.

Hypatia of Alexandria

13

Jessica Manvell

A man, Theon,
obsessed with creating
the perfect human being.
A daughter, Hypatia,
born 370 A.D.
Art, philosophy, literature, and science
all taught to a little girl.
A girl.
Trained in speech, taught in Alexandria,
sent to Athens to study.

Traveling through Europe,
gliding in and out of
lives and minds,
offering glimpses of beauty
and tastes of brilliance.
Proposed to by princes and philosophers,
she had one answer:
"I am already married to the truth."
Thought and discovery were her life,
not submissive silence
in the presence of a man.

She could not accept marriage
or loyalty to a religion.
Questioning and analyzing interfered,
beliefs did not come easily
to this woman of such intellect.
Neoplatonism was her answer,
allowed her her questions,
her doubts,
allowed her an open mind.
In a time when religion was beginning its reign,
Hypatia believed only in possibilities.

Returning to her home,
she found marble fragments
and shards of ivory in ruins
at her feet.
The Christians' wild destruction
had left the University only
a shadow of its earlier days.
Still, Hypatia was determined.
Geometry,
Astronomy,

and the new algebra
consumed her love and devotion.
When her students struggled,
she wrote a commentary on Diophantus.
Then treatises,
And with her father, commentary on Euclid.
From Hypatia also came
A method to distill sea water,
an astrolabe to aid navigation,
a planisphere to chart the stars.

In Alexandria, Hypatia flourished;
her home an intellectual center,
her lectures well attended,
her mind a constant river of ideas.
Her students were in awe,
in love with her,
in love with her knowledge.
They exchanged ideas, called her a
genius,
an oracle,
the Muse.
For her beauty,
her intelligence,
her contributions,
and her spirit of learning,
she was respected and revered.

She was also hated.

Cyril, Christian patriarch of Alexandria,
was the enemy.
Orestes and Cyrene,
friends of Hypatia,
begged her to stay out of danger.

One day in 415
a carriage was stopped on
its way to the university.
An angry mob dragged Hypatia out.
She had become a pawn amidst
anger and the conflict of faiths.

On that day, Hypatia fell.
Her flesh was scraped by oyster shells,
limbs torn from her strong and pure body,
pieces thrown into a fire.
Amidst the victorious chants
of close-minded zealots,
the first woman of mathematics
disappeared
into flames and smoke.

Role Models and Real-Life Experiences

Influencing Girls' Career Choices in Mathematics and Science

14

Edith M. Kort

Women continue to be underrepresented in careers related to mathematics. In industry, only about 12 percent of the scientific and engineering labor force are women (National Research Council 1994). There are some hopeful signs. Some other careers related to mathematics do not have as great a gender disparity, and some studies show that the number of women in careers related to mathematics is increasing. However, there are still fewer women than men in many of these mathematics-related careers, and the gender disparity is greater at top management levels.

To increase the number of women in careers related to mathematics, we must interest girls in these careers and encourage them to study the necessary mathematics. This encouragement should start in early elementary school, since girls as young as third grade lose interest in mathematics (Clewell, Anderson, and Thorpe 1992). Girls must see that careers in mathematics and science are accessible, girls must have experiences that spark their interest in these fields, and girls must meet role models who can help them set their own goals for scientific careers.

ROLE MODELS

Surveys conducted by the Saskatchewan School Trustees Association (SSTA) in May 1994 found that several factors influenced girls' career choices. Parental expectations and the opinions of older siblings strongly affected career choices, but role models and nontraditional experiences were also significant factors (SSTA 1994). A study at Arizona State University reinforces the notion that role models early in a girl's life are an important influence on her aspirations to seek a career in science (Moffat et al. 1992). Recommendations from Karen Hanson of the Education Development Center include introducing children to a variety of career options early and encouraging young women to participate in extracurricular activities that involve mathematics and science (Hanson 1992)

At the University of Rochester Math, Science and Computer Camp, we introduce girls from eight through twelve to a variety of careers in mathematics, science, and technology. The camp is sponsored by the Warner Graduate School of Education and Human Development, the Susan B. Anthony Women's Studies Program, and the Susan B. Anthony University Center at the University of Rochester. The camp is a two-week, all-day program for twenty-four girls. The girls come from various backgrounds: some girls attend private school, some girls attend city schools (many of these girls receive scholarships), and most girls attend suburban public schools. The girls are fairly evenly distributed within the age range and represent the dominant ethnic groups in Rochester. The camp program is staffed by a director and an assistant director, both experienced educators, as well as two high school student assistants.

At camp we promote interest in careers by offering girls opportunities to meet role models, to visit scientists in their laboratories, to do activities with real mathematics and science, and to learn about famous women scientists and mathematicians. Every day we either go on a field trip to a laboratory on campus or a guest speaker comes to camp. We ask guest presenters to share information about their background, explain and demonstrate their career work, and have a hands-on activity for the girls. Making connections among sciences and activities is another objective of the program. Mathematical connections are an important aspect of the problem-solving process (NCTM 1989). Through the visits, the activities, and the computer work, we seek to build girls' self-confidence in mathematics and science, and we try to influence them to study more mathematics and science and set career goals in these fields.

In the July 1995 camp session, we were particularly successful in having a wide range of role models who gave presentations (see fig. 14.1). The goal was to have one guest speaker every day. Some speakers were from the university community; some were from the greater Rochester business community. There was a good mix in the variety of role models who gave presentations at camp: high school students, undergraduate students, graduate students, and established professionals.

The girls are also introduced to role models through a biographical theme, which we present on the first day of camp. In 1995 we used a theme of women inventors. In other years we have used women mathematicians, women scientists, and women Nobel Prize winners in science. We display pictures along with a one-page biography of the women. The girls read the biographies in pairs so that they can discuss and clarify meaning (Siegel et al. 1996). The girls introduce the role models to the entire group of campers by giving a short summary of the biography. Using this biographical information, the girls, who work in two separate groups for some activities, choose names for each group. In 1995 we had the "Pennington Potts" (after Mary Engle Pennington, who developed cooling systems for railroad freight cars, and Mary Florence Potts, who designed a double-pointed iron with a detachable insulated handle) and the "Funky Foxes" (after Sally Vreseis Fox, who developed naturally colored cotton fabrics) (National Women's History Project 1993). After two weeks of calling one another by these women's names, they can easily answer the question, "Are there any women inventors?"

Real Experiences with Role Models

We seek to make the experiences with mathematics and science realistic and resemble those in the real world of work. In school, many mathematical experiences consist of artificial problems that are constructed to teach a mathematical technique. Such problems often bore students. At camp, the girls become interested in, and gain a greater appreciation of, a career when they engage in a hands-on activity related to that career. They also understand the career with greater clarity, make more connections, and remember the activity and career more clearly. The *Professional Standards for Teaching Mathematics* advocates that students should encounter mathematical ideas in a context of genuine problems and situations (NCTM 1991, p. 19). With the hands-on activities led by guest presenters, the girls work on realistic problems from a variety of mathematical and scientific careers. The National Research Council (1989) states that students "must engage mathematics as a human activity" (p. 61). At camp, the girls do mathematics in situations that the presenters encounter as part of their work. Often the girls do not initially recognize the activities as mathematical because they see the context, which does not resemble school mathematics (Kort 1996). At the end of camp, girls have approached me asking, "Why is this

Level	Speciality	Gender	Description of Experience	Type/Length
High School Students	Physics	female	The high school girls in a PRe-college Experience in Physics (PREP) at the university held a science fair for the campers. The Campers did experiments with Newtonian physics which were designed by the PREP campers.	Hands-on 2 hours
Undergraduate College Students	Psychology	female	Sandra conducted language-learning experiments with two campers. She later presented the study and results of her work to the campers.	Lecture 20 minutes
	Biology	female	Aubrae came to camp and described her experiments with fruit flies.	Lecture 15 minutes
	DNA Research	female	Elizabeth described a cell and explained her research with cells and how DNA works.	Lecture 20 minutes
	Chemical Engineering	female	The campers did a variety of experiments that are normally done in the first-year engineering course. They worked with switches on the computer to explore binary numbers, used temperature probes and graphed the results, and made color separations of marker ink with the assistance of undergraduate students. Two undergraduates presented their research projects. The students then did several experiments with the equipment in the laboratories.	Hands-on with a short lecture 2.5 hours
Graduate Students	Laser Lab	female	Brooke and Lily gave the girls a presentation on lasers and how the highest-powered laser in the world works. "T," an established professional, gave a tour of the viewing gallery and explained how a "shot" is taken.	Lecture and tour 1 hour
	Earth and Environmental Science	female/ male	Robin helped the girls to X-ray a rock and use the computer to identify the rock from its graphed X-ray characteristics. Melissa and David explained the geological time line and let us handle fossils.	Hands-on with a short lecture 1 hour
Established Professionals	Optics	male	Greg explained a display of holograms and described how holograms are made. One girl made a hologram.	Lecture and hands-on 45 minutes
	Nursing	female/ male	The girls visited the nursing center's facilities. They used stethoscopes, blood pressure cuffs, and rehabilitation equipment. They then viewed a video on careers in nursing.	Hands-on and a video 1 hour
	Video Photography	female	Jane, the director of a local cable TV station, presented guidelines for making a commercial that the girls then designed. She brought video cameras, and the girls taped interviews with one another.	Hands-on with a short lecture 1 hour
	Computers	male	Bob, Nick, and Dave, the hosts of Sound Bytes, a radio call-in show about computers, gave the girls the opportunity to ask their own questions about computers.	Lectures, question & answer 1 hour
	Zoo Careers	female	Lynn spoke about the variety of careers at the zoo and brought animal artifacts for the girls to explore.	Lecture and hands-on 1 hour
	Electrical Engineering	female	Lynn and Cindy brought in cameras and color separations for the girls to use. She also explained how light is used to make pictures and had the girls model how photons work.	Hands-on and lecture 1 hour
	Dietetics	female	Sandra brought foods to make fruit snacks, spoke about the food pyramid, and talked about nutrition and food labeling.	Lecture and hands-on 1 hour

Fig. 14.1. Role models present their work

called Math Camp when we didn't do any mathematics?" When I ask them to think about the activities that we did, and ask them to think where mathematics was used, they are able to answer their own question.

Students should be involved as active learners with many hands-on activities (Hanson 1992). At camp, the primary method of learning is through active learning and real experiences. Asking the guest presenters to have a hands-on activity for the girls has been successful both for the campers and the presenters. Many of our presenters are used to addressing adult audiences. When we suggest that they include a hands-on activity, they often find that it helps them prepare a more effective presentation and keeps the girls from getting bored with a one-hour lecture.

Most speakers give a short lecture followed by a hands-on activity designed to show the girls what their work is like in the real world. This format has been successful at camp over the years.

In one activity, the girls worked with the director of a local cable television station to learn how shows are made. We invited the director because she runs a high-quality production and television station. She uses current technology and innovative ideas and is a strong role model for the girls at camp. She presented an overview of how a commercial is made, and then she assigned groups of girls to design a commercial, after considering budget constraints. Each of the two groups consisted of advertising designers, whose task was to create the commercial, and business executives, who were the purchasers of the commercial. The advertising designers had to give a presentation to the business executives to sell them the commercial. Each group spent time planning the advertising campaign. Then the girls videotaped interviews by using two video camcorders that the director brought along. The girls had the opportunity to role play and to receive expert advice on the business of cost constraints. Although the scope of the project was large and not fully implemented, the girls had a chance to use mathematics ideas (budget), to problem solve, and to operate technical equipment.

Another activity was an interaction with a nonstandard group of role models. The university hosted a summer program for high school girls, the PRe-college Experience in Physics (PREP). The PREP girls held a science fair for our math, science, and computer campers and developed eight experiments in physics for the math, science and computer campers to do. This experience was unique because girls taught other girls; it gave the math, science, and computer campers a chance to interact closely with role models who were only slightly older than themselves. It was clearly a learning experience for the math campers, but it was also a learning experience for the PREP students. The PREP students had to understand their research project well enough to explain it to others, they had to be prepared for unrehearsed questions, and they had to reflect on their knowledge of the research in preparing the presentation and in making a poster summarizing the research.

A third experience was the interaction with Sandra Barrueco, an undergraduate psychology student who was a visiting researcher at the university for the summer. She asked the two youngest campers to participate in her research. We were glad to accommodate her (with parental permission) and countered with a request that she present her research after she collected the data and analyzed the results. Although only two campers had the hands-on experience of participating in the experiment, all campers learned about the research. Sandra's experiment focused on language learning. Each camper had a session during which she used the computer to draw pictures while a tape of sounds played in the background. After each session of drawing and listening, Sandra tested them to see what sounds they recognized from the tape. After Sandra presented her research to the campers, the experiment's subjects made a few comments. This was a special experience because two of the campers were part of the research experience and then played a role in the presentation.

Our visit to the chemical engineering department was another unusual experience. The girls participated in activities from the first-year college chemical engineering course. The chair of the department and two women students presented the activities. Each camper sat at a computer that had a panel of eight pairs of lights and switches connected to the computer. When the girls keyed in a number, a sequence of lights turned on. This sequence of lights corresponded to the binary value of the number. Some of the girls noticed patterns between the lights and the numbers. As a second part of the experiment, the procedure was reversed. The girls turned the lights on with the switches, and the computer then converted their binary number to decimal and displayed the decimal equivalent. With some guidance about the meaning of the lights and switches, the girls began to understand binary numbers. Next, they worked with temperature probes attached to the computers. Changes in the temperature were graphed on the computer screen. The girls saw the computer as a device that could sense switch settings and interpret them as numbers, that could receive keyboard input and give output to the switch lights, and that could sense temperature and graph it.

After the initial hands-on work, two other women students gave short presentations on their summer research. One brought her research log and explained how she used it, which made the connection between the scientist's daily writing of logs and the girls' daily writing in journals. The campers then visited several laboratories in which graduate students demonstrated experiments that required specialized equipment. This field trip provided a comprehensive perspective on chemical engineering, given the variety of role models and activities.

Real Experiences at Camp

Experiences at camp that resemble those of scientists in the real world are not limited to those with guest presenters. During an eight-hour day, the girls do a number of mathematics and science activities in classrooms, at lunch and snack time, during field trips, and at the computer.

For example, the girls did the following series of activities with polyominoes, working first with tetrominoes and then with pentominoes.

1. Find the tetrominoes—all the possible combinations of four squares. The girls drew combinations on centimeter graph paper and compared answers.
2. Play Castle Calibration, a pencil-and-paper game in which they use the tetrominoes to complete a picture.
3. Play Tetris (Pajitnov 1985) on the computer. In Tetris, a random series of tetrominoes falls one at a time from the top of the screen and fills a well; the goal is to fill rows without any gaps. This game gave the girls a different perspective on combinations that form tetrominoes.
4. Draw the pentominoes (combinations of five squares).
5. Decorate and cut out a set of pentominoes. The girls became familiar with the distinct shape of each pentomino.
6. Form a rectangle by using all the pentomino pieces.

The girls worked with polyominoes without thinking that they were doing mathematics. They had a variety of experiences with spatial visualization, an area in which girls may not perform as well as boys (Tartre 1990). The girls worked on open-ended problems with multiple solutions. The activities were designed to move from simpler problems to related problems that were more complex, following a scaffolding model (Collins, Hawkins, and Carver 1991).

In addition to numerous activities using M&Ms (distribution, estimation, comparison, and so on) and nutritional labels, we held a watermelon-seed spitting contest. Each girl ate a wedge of watermelon and counted the seeds. They looked at the number of white and black seeds and compared size and number. They saved three seeds for spitting; they measured the one that went the farthest. They completed a table with an estimate and a measurement of the farthest seed, both in standard units and in nonstandard units of their choice, such as a pencil length, a Lego block, or a finger. It was clear without measuring which seed traveled the farthest. They made the standard measurement with a standard ruler, so the girls had to develop strategies to measure distances that were longer than a ruler. They also had to ensure that they were measuring a straight line so that they did not add extra distance to the measurement. As in the previous activity, we integrated a number of mathematical ideas into the project, yet made it fun.

We started each camp day with a warm-up brain-teaser puzzle. Some came from *Games* magazine and related publications. Others were puzzles from problem-solving books. We sometimes found a puzzle related to another activity, such as a problem-solving activity with rotating objects before we worked with tessellations. The introductory activity served to activate prior knowledge, thus forming a stronger foundation for further learning and helping make connections.

Although these experiences took place in what may be a privileged setting—at a university with interested students, accessibility to laboratories, and no curriculum or time constraints—experiences with role models and real-life settings can happen in other environments. Many of these field trips and guest speakers can be part of a mathematics and science club at school. They can even be enrichment activities in a classroom.

The value of real experience with role models cannot be underestimated. One parent commented to me at the end of camp, "Every day my daughter came home wanting a different career." Because about 25 percent of girls return to camp for a subsequent year; the activities each year are different. Teachers and parents have told us about the contributions that the girls were able to make in school on the basis of their camp experiences. With the opportunity to do activities in context and the interaction with hands-on activities, the girls gain a greater perspective on the variety of careers in science.

REFERENCES

Clewell, Beatriz C., Bernice T. Anderson, and Margaret E. Thorpe. *Breaking the Barriers: Helping Female and Minority Students Succeed in Mathematics and Science.* San Francisco: Jossey-Bass, 1992.

Collins, Allan, Jan Hawkins, and Sharon M. Carver. "A Cognitive Apprenticeship for Disadvantaged Students." In *Teaching Advanced Skills to At-Risk Students: Views from Research and Practice,* edited by Barbara Means, Carol Chelemer, and Michael S. Knapp. San Francisco: Jossey-Bass, 1991.

Hanson, Katherine. *Teaching Mathematics Effectively and Equitably to Females: Trends and Issues,* No. 17; ERIC Clearinghouse on Urban Education; Institute for Minority Education, July 1992.

Kort, Edith M. "Expanding the Horizons of Young Women with Worthwhile Mathematical Tasks." *Focus on Learning Problems in Mathematics.* 18, nos 1, 2, and 3 (1996).

Moffat, Nancy, Michael Piburn, Larry P. Sidlik, Dale R. Baker, and Rick Trammel. "Girls and Science Careers: Positive Attitudes Are Not Enough." Presentation at the annual meeting of the National Association for Research in Science Teaching, Boston, Mass., March 1992.

National Council of Teachers of Mathematics. *Curriculum and Evaluation Standards for School Mathematics.* Reston, Va.: National Council of Teachers of Mathematics, 1989.

————. *Professional Standards for Teaching Mathematics.* Reston, Va.: National Council of Teachers of Mathematics, 1991.

National Research Council. *Everybody Counts: A Report to the Nation on the Future of Mathematics Education.* Washington, D.C.: National Academy Press, 1989.

————. *Women Scientists and Engineers Employed in Industry: Why So Few?* Washington, D.C.: National Academy Press, 1994.

National Women's History Project. *Inventive Women* (poster series). Windsor, Calif.: National Women's History Project, 1993.

Pajitnov, Alexey. Tetris. San Francisco: Blue Planet Software, 1985. [software].

Saskatchewan School Trustees Association. *Female Student Career Aspirations in Science.* Saskatchewan School Trustees Association Research Centre Report: #94-04. Sasketoon, Sask.: Saskatchewan School Trustees Association, 1994.

Siegel, Marjorie, Raffaella Borasi, Judith M. Fonzi, Lisa Grasso Sandvidge, and Constance Smith. *Using Reading to Construct Mathematical Meaning.* Reston, Va.: National Council of Teachers of Mathematics, 1996.

Tartre, Lindsay. "Spatial Skills, Gender and Mathematics." In *Mathematics and Gender*, edited by Elizabeth Fennema and Gilah C. Leder, pp. 27–59. New York: Teachers College Press, 1991.

Remarkable Women of Mathematics and Science

15

Leah P. McCoy

One of the major issues that contributes to gender inequity in mathematics is the perception of mathematics as a male domain. Too often, the only people mentioned in mathematics classes are male mathematicians and scientists. Because female students are not aware of female mathematicians and scientists, they may internalize a belief that mathematics is not appropriate for women. As educators, we can help these students form more positive attitudes toward mathematics by presenting female role models who have been successful in mathematics. By personalizing these historical figures, we may also reduce mathematics anxiety as the people who do mathematics become more real. We need to make the study of remarkable women an integral part of the K–12 mathematics curriculum. The following activities are suggested for the middle school level, but they are easily adaptable for grades four through twelve.

THE REMARKABLE WOMEN ACTIVITY

For the Remarkable Women activity, students use Internet resources to research the history of mathematics and women in mathematics. This activity may also be a part of the computer curriculum because students need to know how to use the World Wide Web for information retrieval. Individual worksheets guide the study of each remarkable woman. Sample worksheets for middle school students are included as figure 15.1 (Sophie Germain), figure 15.2 (Caroline Hershel), figure 15.3 (Grace Murray Hopper), and figure 15.4 (Hypatia). Each worksheet contains information and questions about the woman's life and connections to her area of study and work. The material was developed through searches on the Net. Each worksheet suggests relevant Web sites, which students can search for further information. In addition to completing the short-answer questions, students write another interesting fact about the remarkable woman and give their thoughts about pursuing a possible future career in her discipline. If students work in pairs or in groups of three, they will benefit from the interaction as they answer these questions.

To study other remarkable women, the teacher can use this format to create additional worksheets, or small groups of students can work together to create worksheets to exchange with other groups. In addition to the obvious gains in historical information and mathematical connections, students enjoy the activity and their self-esteem improves as they gain computer skills.

News Release. Students research a mathematician and write a news release announcing an important event in her life. Have them pretend that the event has just happened and tell all important details. Video equipment can be used to tape and present these "newscasts." This is an excellent exercise for developing communication skills, and students enjoy dressing as newscasters and presenting the news. See figures 15.5 and 15.6 for examples of news releases.

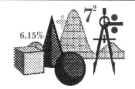 

Sophie Germain
Mathematics

(1776–1831), France

Education: She received no formal education. Women were not allowed to attend school. She was to receive an honorary degree from the University of Göttengen, but died before it could be awarded.

Experience: She studied mathematics independently and was given several awards for her work, which she submitted in men's names.

WWW Information Sites (You may also want to do a NET SEARCH):
   www.scottlan.edu/lriddle/women/germain.htm

Fact 1: Because women were not allowed to study at the university, Sophie Germain obtained mathematics notes from male friends and submitted work in what name? _____

Fact 2: Sophie Germain is best known for her work in what area of mathematics? _____

_____

Another interesting fact about her: _____

_____

CONNECTION: Prime Numbers

Sophie Germain studied various kinds of numbers, including prime numbers. A prime number is an integer greater than 1 whose only positive divisors (factors) are 1 and itself.

WWW Information Sites (You may also want to do a NET SEARCH):
   www.utm.edu/research/primes/largest.html

Q1: What are the prime divisors of 10? _____

Q2: How is the Sieve of Eratosthenes used? _____

Q3: What is a Sophie Germain prime? _____

My thoughts about a career in mathematics research: _____

_____

_____

Fig. 15.1. Remarkable Women worksheet: Sophie Germain

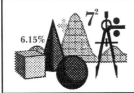

Remarkable Women 2

Caroline Herschel
Astronomy

(1740–1840), Germany, England

<u>Education:</u>   She was not allowed a formal education, but was taught by her brother, William.

<u>Experience:</u>   She worked with her brother to study the heavens.

WWW Information Sites (You may also want to do a NET SEARCH):
www-groups.dcs.st-and.ac.uk/~history/Mathematicians/Herschel_Caroline.html

Fact 1: What was Caroline Hershel's main job in helping her astronomer brother William?

_____

Fact 2: For what work was she awarded a gold medal by the Astronomical Society in 1828?

_____

Another interesting fact about her: _____

_____

CONNECTION: Solar System

Caroline Herschel studied heavenly bodies, including the solar system.

WWW Information Sites (You may also want to do a NET SEARCH):
www.windows.umich.edu

Q1: What are the planets of the solar system?   _____

_____

Q2: What planet was discovered most recently, and when was that?   _____

Q3: What are small, dense objects orbiting the Sun called?   _____

Q4: What are smal, icy objects with highly eccentric orbits called?   _____

My thoughts about a career in astronomy: _____

_____

_____

_____

Fig. 15.2. Remarkable Women worksheet: Caroline Herschel

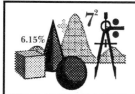

Remarkable Women 3

Grace Murray Hopper
Computer Science

(1906–1992), United States

Education: B.A., Vassar College
M.A. and Ph.D. in Mathematics and Physics, Yale University

Experience: She taught at Vassar College and Harvard University.
She served in the U.S. Navy and retired in 1986 as a rear admiral.

WWW Information Sites (You may also want to do a NET SEARCH):
web.www.mit.edu/invent/www/inventorsA-H/hopper.html

Fact 1: What important component of computer systems did Grace Hopper invent? _____

_____

Fact 2: What is the name of the higher-level programming language, which Grace Hopper developed? _____

_____

Another interesting fact about her: _____

_____

CONNECTION: Computers

Grace Hopper was a significant pioneer in the development of computers.

WWW Information Sites (You may also want to do a NET SEARCH):
www.digitalcentury.com/encyclo/update/comp_hd.html

Q1: What is the name of the first computer, which emerged about 5000 years ago in Asia Minor and is still in use today? _____

Q2: What was invented in 1948 that greatly changed the development of the computer by allowing it to be much smaller? _____

Q3: What "computer development" did U.S. Vice President Al Gore promise to make a priority in 1992? _____

My thoughts about a career in computer science: _____

_____

_____

Fig. 15.3. Remarkable Women worksheet: Grace Murray Hopper

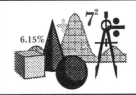

Remarkable Women 4

Hypatia
Mathematics

(370–415 A.D.), Egypt

Education: Her father, a mathematics professor, taught her.

Experience: She was a professor at the University at Alexandria.

WWW Information Sites (You may also want to do a NET SEARCH):
www.cosmopolis.com/people/hypatia.html

Fact 1: Hypatia's studies included mathematics and what two other subjects? _____

_____

Fact 2: Why was Hypatia killed by a mob? _____

_____

Another interesting fact about her: _____

_____

CONNECTION: Fractals

Hypatia studied cones and other geometric figures. Currently, many mathematicians are studying geometric figures called *fractals*.

WWW Information Sites (You may also want to do a NET SEARCH):
cml.rice.edu:80/~lanius/frac/

Q1: What do fractals often look like? _____

Q2: What is one famous fractal? Sketch it

Q3: What are the three properties that define a fractal? _____

My thoughts about a career as a mathematics professor: _____

_____

_____

Fig. 15.4. Remarkable Women worksheet: Hypatia

NEWS RELEASE

1 June 1933

EMMY NOETHER TAKES POSITION AT BRYN MAWR COLLEGE

Emmy Noether, a noted mathematician, has accepted a position as a visiting professor of mathematics at Bryn Mawr College near Philadelphia, Pennsylvania.

Ms. Noether was born in 1882. She grew up in Erlangen, Germany, and learned mathematics from her father, who was also a mathematics professor. She earned a doctorate at the University of Erlangen in 1907. She had difficulty getting a position as a professor because all such positions were held only by men. She recently held a position as a lecturer at the University of Göttingen.

Because of her Jewish heritage, Ms. Noether was forced by the Nazi government to resign from her university position and leave Germany.

At Bryn Mawr, she will teach mathematics courses and continue her research in algebra. She also expects to work with Albert Einstein.

Fig. 15.5. News Release: Emmy Noether

NEWS RELEASE

16 November 1845

ADA BYRON LOVELACE INVENTS COMPUTER PROGRAMMING

Ada Byron Lovelace, working with the Difference Engine idea with Mr. Charles Babbage, has invented a way to make the machine respond to a series of instructions. She used punched cards to enable the machine to play music.

Countess Lovelace, who was tutored by another famous woman mathematician, Mary Fairfax Somerville, has worked with Mr. Babbage for many years. She is responsible for codeveloping and publishing much of the work on the Difference Engine and the Analytical engine.

Her invention of programming, or teaching the computer, will make the computer an important tool for the future.

Fig. 15.6. News Release: Ada Byron Lovelace

Poster Fair. Students research a mathematician and prepare a poster that presents her achievements. This activity appeals to the artistic student, and students can use various materials to construct the poster. Posters may be displayed in a hallway, cafeteria, or media center to share information with other students. See figure 15.7.

Math People Notebook. Students keep a class notebook that contains short biographies of mathematicians. They research a past or a current mathematician (or both) and prepare a one-page biography, which they place in a binder in the classroom. Students maintain the notebook throughout the school year, and frequently add and refer to it. See figures 15.8 and 15.9.

We know that countless small actions build a cumulative antimathematics attitude in many girls. To counter this, we must consciously provide many small actions to promote a positive attitude toward mathematics. These middle school activities are useful starting points. Using the suggested references and other resources, teachers and students should include additional mathematicians and scientists in these activities. It is crucial that students be exposed to models of successful male and female mathematicians. In this way, we can encourage all students in their study of mathematics.

Sonya Kovalevskaya (1850-1891)

She was Russian.

Her parents wallpapered her room with pages from a mathematics book.

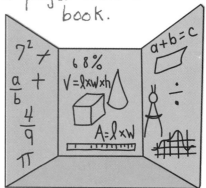

Because women could not study mathematics in Russia...

...she married so she could go to Germany to study math.

She won a great prize for her mathematics research

from the Paris Academy of Sciences in 1888.

She was a professor of mathematics at the University of Stockholm in Sweden.

by Linda K. Wakeley

Fig. 15.7. Poster Fair: Sonya Kovalevskaya

MATH PEOPLE – Maria Agnesi

Maria Agnesi was born in Milan, Italy, on 16 May 1718 to a wealthy and literate family. Her father was a professor of mathematics, and she developed an interest in mathematics early in her life. She was very smart and was considered a child prodigy. She spent her teenage years doing private study and tutoring her brothers. She often participated in her father's seminars, and she published a collection of essays on science and philosophy.

Later, she also published a comprehensive calculus textbook, which was a great success. In this text, Maria discussed a sine curve. Because of a translation error, the curve became known as a "witch." Maria then became known as "the witch of Agnesi."

She lectured at the University of Bologna and was elected to the Bologna Academy of Sciences. In 1762, she gave up mathematics and spent the rest of her life in charity work with the sick and the poor.

Maria is still remembered today. A street and a school in Milan bear her name. Many scholarships are given in her name.

Fig. 15.8. Math People: Maria Agnesi

MATH PEOPLE — Mae Jemison

Mae Jemison was born in 1956 in Alabama. Her family moved to Chicago, where she completed high school in 1973. She attended Stanford University, graduating in 1977 with a double major in chemical engineering and African American studies. She received her M.D. degree from Cornell University in 1981.

After her medical internship in Los Angeles, Mae Jemison joined the Peace Corps and served as a medical officer in Sierra Leone and Liberia.

In 1987, she was accepted by NASA as an astronaut. On 12 September 1992, Mae Jemison took off in the Shuttle *Endeavor*, where she served as a science specialist. Thus, she became the first African American woman in space.

In 1993, she resigned from NASA to work toward providing equity and opportunity for minorities and all people.

Fig. 15.9. Math People: Mae Jemison

SOURCES OF BIOGRAPHIES FOR WOMEN IN MATHEMATICS AND SCIENCE

Cooney, Miram P., ed. *Celebrating Women in Mathematics and Science*. Reston, Va.: National Council of Teachers of Mathematics, 1996.

Osen, Lynn M. *Women in Mathematics*. Cambridge, Mass.: Massachusetts Institute of Technology Press, 1974.

Perl, Teri. *Math Equals: Biographies of Women Mathematicians and Related Activities*. Menlo Park, Calif.: Addison-Wesley, 1978.

Perl, Teri, and Joan Manning. *Women, Numbers, and Dreams*. Santa Rosa, Calif.: National Women's History Project, 1985.

Reimer, Luetta, and Wilburt Reimer. *Mathematicians Are People, Too*. Palo Alto, Calif.: Dale Seymour, 1990.

——. *Mathematicians Are People, Too: Vol. 2*. Palo Alto, Calif.: Dale Seymour, 1995.

Smith, Sanderson M. *Agnesi to Zeno: Over 100 Vignettes from the History of Math*. Berkeley, Calif.: Key Curriculum Press, 1996.

Warren, Rebecca L., and Mary H. Thompson. *The Scientist within You*. Eugene, Ore.: ACI Publishing, 1994.

——. *The Scientist within You: Vol. 2*. Eugene, Ore.: ACI Publishing, 1995.

Note: Figures include students' work from methods courses at Wake Forest University.

Girls-Only Classes in Public Schools

Ambivalence and Support

16

Janice Streitmatter

Heather A. Blair

Joanna Marasco

Girls-only programs in public schools are a recent phenomenon designed to provide an equitable education for girls. However, a great deal of controversy surrounds these classes. This chapter reflects briefly on the effects of public policy and legislation based on frameworks of equality and equity and examines the implications for single-sex schooling in public schools. This discussion sheds light on the disjuncture between public policy, framed by equality, and the practice of girls-only programming, framed by equity. We also discuss the evident ambivalence between the macro levels of policy and the micro, or local, levels of policy and how this ambivalence plays out in the single-sex class of this case study.

THE ORIGINS OF SEX-EQUITY POLICY

The term *equity* is increasingly used to mean "treatment that is fair to women both in form and result" (Jacklin 1981, p. 56). The origin and meaning of the term help us understand single-sex schooling in the United States today. The concept of fairness allows an interpretation beyond that of traditional laws and builds a framework for a broader analysis of the legal status of women (Jacklin 1981).

The premise of equity in teaching is understood as the systematic delivery of a greater-than-equal share of resources to a group of students recognized as at-risk to compensate for historical bias. The framework of equality, in contrast, suggests that within the process—life in classrooms in this example—all students receive the same quality and quantity of resources. An equity approach focuses on the final outcome, usually either achievement or self-esteem. For example, an equity framework theoretically will give the previously disadvantaged group more during the process to ensure that outcomes by group are indistinguishable. The equality approach places emphasizes the process, with the expectation that outcomes will vary by individual, not by identifiable group.

Many who cite the limitations of the construct of equality believe that it does not go far enough to supply the less advantaged with the tools to take part with the advantaged. The equality argument, according to Greene (1985, p. 39), does

> not allow for acknowledgment of the difficulty experienced by female students who cannot perceive themselves as chemists, school administrators, foundation executives, and the like, no matter what formal criteria exist.

Federal legislation and judicial decisions are the main determinates of all public policy. The most prominent and significant policy standard regarding sex in the past three decades is Title IX of the Educational Amendments Act, 1972. A landmark in terms of educational policy and implications for women, Title IX was designed to "prohibit discrimination on the basis of sex in education, activities and programs that receive federal financial assistance" (Klein 1985, p. xii). Before 1972, legal and educational policies toward women ranged from "protecting women," in the 1800s and early 1900s, to providing equality for men and women.

Subsequently, Congress passed the Women's Educational Equity Act (WEEA) in combination with Title IV of the Civil Rights Act in 1974 and the Vocational Educational Amendment (VEA) in 1976. Both support sex equality in terms of educational outcomes and benefits. Since that time, nearly one-third of the states have passed similar legislation (Stromquist 1993).

Title IX has primarily affected the educational system in the areas of access to schooling, school textbooks, curriculum content, teacher preparation, women in administration, and incentives and support measures for girls (Stromquist 1993). Access is the one area in which gains have been obvious. In contrast, other aspects of the educational system affected by Title IX have had varying results. School textbooks and curriculum in mathematics, for example, have been slow to reflect greater gender inclusion.

Girls-only programs in public schools are rarely explicitly discussed in the educational-policy literature. They could be viewed as falling into the category of incentives and support measures, but there is little mention in the literature of girls-only programs in the United States within the context of equity advancement. In terms of Title IX, single-sex classes within public coeducational schools actually contravene this Act. Title IX explicitly allows single-sex public schools, except vocational schools, but not separate classes for boys or girls within coeducational schools.

Title IX has primarily affected the educational system in the areas of access to schooling, school textbooks, curriculum content, teacher preparation, women in administration, and incentives and support measures of girls (Stromquist 1993). Access is the one area in which gains have been obvious.

The judicial system has played an important policy role. Issues of discrimination on the basis of sex have been brought to the court and validated by numerous court decisions. For example, the recent U.S. Supreme Court cases involving the Citadel and the Virginia Military Institute (VMI)(Faludi 1994) reflect the courts' role, and in some cases, interpretation of Title IX as a means to ensure equal access by sex. These male-only academies, some of the last bastions of male privilege, were challenged in the courts by a woman's application for admission. The federal court decision in April 1995 allowed Shannon Faulkner to attend the Citadel as the first female cadet. Other decisions have supported an interpretation of Title IX as a measure of equity. The policy of the Arizona Interscholastic Association, supported by court action, allows female secondary school athletes access to interscholastic football and wrestling teams but does not allow boys access to girls' volleyball or softball teams. A similar interpretation was delivered by the court when a boy sought admissin to Girls High in Philadelphia. Although earlier the court reversed the long-standing tradition of male-only admission to Central High in Philadelphia, they retained female-only admission to Girls High. In these instances, the courts have determined that access by sex is an issue of equity; what is fair and just, not necessarily what is equal.

As illustrated by Title IX and recent Supreme Court decisions, certain degrees of ambivalence exist within the realm of federal policymaking bodies. We believe this ambivalence is manifest in public policy at the local level and is evident in educational practices of the Valley School District and the Britt Murray School.

GIRLS-ONLY MATHEMATICS: A CASE STUDY

This study took place in a girls-only mathematics class in a public coeducational middle school in a southwestern city in the United States during the 1993 through 1996 school years. Britt Murray School is a mathematics and science magnet school in a middle-class neighborhood on the outskirts of the city. The school offers a wide range of programming but focuses on science and mathematics. Some students are bused to Britt Murray from all parts of the city as part of the magnet program, whereas others reside in the neighborhood.

The research was based on an ethnographic research design. Students, the teacher, and the principal were interviewed. We observed the girls-only class as well as the teacher's mixed-gender class of the same subject approximately once every ten days during the 1993–94, 1994–95, and 1995–96 school years.

The mathematics program was designed to have young women enroll as seventh graders and move as a cohort with the teacher into a girls-only algebra class the following year. In the first year of the program, the class comprised twenty-four students. At the end of that year, one moved out of the city and one transferred to a mixed-gender class. Three young women joined the class at the beginning of the eighth-grade year.

Fourteen young women were selected for interviewing during the seventh-grade year. We chose five students who were African American, Latina, or Asian American to ensure that we would collect data from students of color. The other young women were chosen randomly. We followed these young women through the eighth-grade year and reinterviewed nine during their first year in high school. The preliminary results reveal some interesting findings about a number of issues within the girls-only class, but we focus here on the policy implications for girls-only programming.

Ambivalence on the Part of the School District

The Valley School District had no explicit gender-equity policy at the time of our study and nothing that appeared to be a coherent, unwritten working policy. However, the district had policies called "Diversity Appreciation Education Policy" and "Gifted and Talented Education Policy." Although neither policy directly addressed gender and equity or single-gender schooling, the first addressed "eradicating sexism" in very general terms. The latter stated that no child would be excluded from the gifted and talented program on the basis of gender, ethnicity, handicapping condition, or religion. Yet the single-gender classroom existed in the school district, and by the third year of our study, two other schools had girls-only programming in either mathematics or science. There was no local policy to support these programs and none to discourage it. However, the programs were not widely advertised. It was as if the school district privately approved but was not prepared to do so publicly. The ambivalence was obvious.

Ambivalence on the Part of the Principal

The principal was responsible for initiating and implementing the girls-only mathematics class in this school in 1992. He had read a newspaper article about a girls-only mathematics class in a public high school and seized the idea as one worth trying in his middle school. He selected the students, appointed the teacher, and informed the rest of the staff.

The principal believed that this enhancement of opportunities would benefit the young women and was proud of his initiative. He was supportive of them and open to questions from the parents. He believed that young women should have an equal chance in mathematics and science and thought that this program would help "his girls," although he had never read any of the academic research literature regarding single-gender schooling. His approach to the girls-only program, as well-meaning as it was, led us to understand that he did not really understand or embrace equity and perhaps not equality.

The principal's actions and the manner in which he undertook this project were indicative of a kind of ambivalence. He did not consult in advance with anyone about developing this project—not his staff, his superiors, or the young women's parents. The students selected for the initial two years were those whom he believed to be relatively high achievers, with some multicultural mix. He was open to the idea of having a research team from the university follow the young women's progress, but he never showed any interest in the findings and never involved himself in any way. Lastly, he did not want the existence of this class to be public knowledge. His actions of keeping the class as secret as possible and not involving the teacher in the decision-making process perpetuated a sense of ambivalence about the importance of the class.

Throughout the three years of the research, the principal's ambivalence toward this program continued to be evident. He started a second girls-only class in the school, seventh-grade science. He selected the teachers for mathematics and science, yet gave them no support or professional training specific to this undertaking; no dialogue between teachers took place. He scheduled girls-only science at the same time as girls-only mathematics so that the same young women could not take both classes. Some parents wanted their daughters in both classes, but he did not see this as appropriate. His ambivalence was evident in his decision making. He made very immediate decisions, many of which were spontaneous and random.

Ambivalence and Certainty on the Part of the Teacher

The teacher started out very uncertain, with her own set of ambivalent attitudes toward this project. Immediately after being informed of her appointment to teach this class, she began to worry. She did not know whether she could or wanted to teach it. At the beginning of the second year of the program, she commented:

> I was afraid it was going to be illegal. I was more worried about that than anything, and then I was afraid that I didn't know enough about gender issues in general to even be successful as it was. And I was new to the school and I was feeling swamped, and I thought "Gosh, they're giving this to me and I don't even know what to do." I was worried about the reactions of my colleagues, which were very negative.

Her ambivalence was reinforced by the response of some of her colleagues in the school. Several of the other mathematics teachers commented to her that they believed that a girls-only classroom was not fair to them because they now had a larger proportion of boys in their sections.

The teacher, however, was a real advocate for the importance of mathematics in the curriculum. She loved her subject matter and wanted to be a good teacher. She was proud that she was one of very few women in her advanced mathematics courses at college and had done as well as the men. She thought that it was wrong that there were so few women in mathematics, but she believed that it was important in a field like mathematics for a woman to show that she was as smart and as competent as the men. Being in a class with men was a way to demonstrate that competence. The teacher was adamant at the beginning of the project that the instruction and content of the girls-only mathematics class would not be any different from that in her other classes. She was determined to treat all her students the same. She believed in the efficacy of an equality approach, fearing that to do otherwise would in some way victimize the girls.

Throughout the first year, she felt ill prepared, on the one hand, and excited by what she was experiencing on the other. She described the girls-only class at the end of the fist year in this way:

> They were more willing to take risks, and they were more willing to be responsible. They'd always excite me; sometimes they would make these jumps that my other classes couldn't or wouldn't make and that always made my day because, you know, that's what you look for. You look for those little leaps that kids make, and they would do it at least once a week if not more, and so I used to feel very good about the class. I thought "OK, this is what's supposed to be happening." They seemed to be much more positive about mathematics than girls in my other classes. They seemed much more goal-oriented than girls in my other classes, and it made me feel like, "Well maybe I was wrong; maybe this is a good thing; maybe even if it's illegal we should still do it."

It seemed the teacher's views of learning and teaching in this classroom also reflected her ambivalence. This ambivalence included her choice of methodologies. She worried that she was spending too much time on helping the girls understand the reasons for the correct answers, thereby running the risk of leaving out some curricular content. However, she believed that the time was well spent because the students were learning the process.

> I thought to myself, "How do I balance how I teach them with what they have to know in order for them to have a defined body of knowledge called algebra?"

However, during the three years of the study, the teacher changed from being ambivalent about the value of this class and her practices to being quite certain of their value—a dramatic change. Throughout the research, she kept asking questions of herself and of the research team. She also started to read on the topics of gender, gender equity, and gender and schools. When she read *Failing at Fairness* (Sadker and Sadker 1994) at the end of the 1994 school year, she commented on how valuable it would have been had she had the chance to read and think about these issues before she started teaching the girls-only class.

Toward the end of the second year, she shared her thoughts about possible differences in her teaching in the girls-only class and in the mixed-gender classes:

> I know now, after I've given it a lot of thought, that I do work with the girls-only group differently than the other mixed groups. I really believed I didn't, but now I know I do. The girls bring out more in me while I try to bring out more in them. I'm just as tough on them, maybe tougher, but it's different. And it's not as if their behavior is any better than the boys and girls in other classes. They're really chattier than other groups. But I know that I don't have to spend energy ignoring some of the really immature behavior of some of the boys. And I don't have to work so hard drawing out the girls. In the girls-only class, they draw themselves out. It's very different. And I know I teach differently too. The questions we get into allow me to concentrate more on process with the girls. I don't really know why.

At the outset, she had operated primarily from a belief in equality; later, she moved to an acceptance of provisional equity, although she still had some reservations. She became able to ignore the force of the principles of equality in the policies all around and the actions of the principal, who himself was ambivalent.

By the end of the three years, even though she applied for a transfer to another school, she was no longer ambivalent. She registered her own daughter, then in the seventh grade in Britt Murray School, in the girls-only mathematics class so that she would have the benefit of this experience. Awareness on the part of this teacher made a difference. The teacher had had many questions. The more she learned the more certain she became that the girls-only mathematics class was the best setting for middle school girls.

The Young Women: No Ambivalence on Their Part

The young women, in contrast, displayed no ambivalence. They were pleased at the absence of boys from the first day; their resolution that this setting was powerful for them as mathematics students only grew. When they watched the *Dateline NBC* television program that reviewed the work of the Sadkers, some of which is found in *Failing at Fairness* (Sadker and Sadker 1994), one said, "They should have interviewed us, we know what it's like." Kendall, an African American eight grader, described how her two years in the program had affected her confidence about mathematics:

> I consider myself a good student, but sometimes I've been afraid to show it. In here it's OK to do well. Guess when I don't feel like I can answer questions in class and show people that I can do it, I start not liking the class and the things we're supposed to learn. But in here I can be myself, feel smart, and get good grades. I'm going to take geometry next year in high school. I just wish that there was a girls-only class for that. But at least I feel like I can do the mathematics. I know I can.

Unlike Kendall, Nona, a Mexican American young woman, did not consider herself a strong student to begin with. But by the end of the eighth-grade year, she described her feeling about herself in mathematics class this way:

> Since the class, I've liked mathematics a lot more than before. I see the usefulness of mathematics in a lot of other things—like I'm into astronomy now, and you need mathematics for that. If I had to start all over, I would take this class instead of the regular ones. There was still some stuff in it that I'm not sure I understood, so I'm going to take algebra next year in high school and then I'll take geometry during the summer so I can catch up. That way I can be ready for calculus. I never thought I would go on in mathematics. In fact, a lot of my friends in the neighborhood talk about dropping out altogether. But I'm not going to do that, and I'm going to take more mathematics. I do wish that there were more girls-only classes, some in high school. I'm not afraid to ask questions and get things wrong. I'll miss being with just girls, but I'll be OK now.

These two data samples reflect the strong convictions of these young women about the importance of their girls-only experiences. It is important to note that every young woman interviewed described this conviction, with some variation in intensity among them. Further, the observational data corroborated much of what the young women talked about, especially regarding their sense of empowerment about feeling free to speak out in class without the fear that others might make fun of them. For example, questions to the teacher, and wrong answers in response to teacher-initiated questions, were constant. Quiet moments were rare, and never once did a student demonstrate a reluctance to speak.

MACROANALYSIS AND MICROANALYSIS COMPARISON

Ironically, legislation intended to address the disparity of outcomes in schools by gender may actually perpetuate this ongoing problem. By enforcing an organizational structure through narrowly interpreted federal policy, insisting on coeducational classes, and forbidding single-sex classes, the education system subverts the goals of Title IX. The contradictory natures of the Title IX legislation and of single-sex classes contribute to the ambivalence on the part of administrators. However, girls-only classes are being developed in a number of locations in the United States, despite their apparent contradiction of Title IX. The principal of the Britt Murray School initiated this project under the assumption that he would improve the opportunities for young women, even though this action violated the law. This situation demonstrates the tension between the intent of public policy at the federal level and practice at the local level.

The limited data from the participants in the girls-only mathematics class suggest that the girls-only mathematics class at Britt Murray School was perceived as beneficial by the principal, the teacher, and especially the young women. Further evidence shows that many parents in the school community supported the girls-only class. Despite the partial cloak of secrecy that the principal attempted to draw over the class's existence, a waiting list for enrollment in the class for the second and third years was created early in the spring of the first year. The central issue, then, appears to be how to resolve this tension between macro policy and micro, or local, practice.

The most fundamental difference seems to lie within the interpretation of what the gender-equitable practices and programs are meant to resolve. If the purpose is to treat gender inequities and the perpetual disempowerment of female students on a short-term and superficial basis, then it would appear that policy and practice within the theoretical framework of equality should continue. This approach defines gender-equitable programs and practices as innovations—quick-fix remedies that with some teacher training can be implemented without making anyone uncomfortable. With innovation, the balance of resource distributions is not disrupted, and the group that historically has received the greater amount of resources continues to do so. However, the assurance of equal access to resources does not ensure equal treatment within the local classroom, nor does it create a place where beliefs and attitudes about potential and capabilities based on gender change. Classrooms remain places where males are dominant and females are likely to feel less than equal. Too much evidence exists that tells us that an equality approach is not sufficiently powerful to ensure equal outcomes free of gender-group disparity.

To varying degrees in our study, school administrators, as well as teachers, students, and their parents, recognized that girls-only classrooms create an important option for female students. In these classrooms, young women come together, not because there is something wrong with them that needs to be repaired, but because there are things that are right and must be further empowered. The development of female collectivity (Walkerdine 1990), where a community is created in a safe environment and where risks can be taken, appears to be a crucial element of the girls-only class (Streitmatter 1997).

Girls-only classes represent a departure from the equality or innovation construct. They reflect a redistribution of resources according to an equity, or reform, approach—the more viable approach if we truly seek change. Teachers who complain that the creation of a girls-only class is "unfair" because they then have classes with "too many boys" or parents who complain that special classes for girls are "unfair" to boys are not correct. A girls-only classroom is unequal but not unfair. As we pointed out earlier, "equity jurisprudence raised fairness above the application of strict rules of precedent ... [and] equity placed fairness above traditional rules of law" (Jacklin 1981, p. 56).

The need to realign resources through the development of girls-only classes must be acknowledged at the federal level. Local voices, especially those of the young women who tell the compelling stories of their experiences in girls-only classes, must be acknowledged as legitimate. With a broader interpretation of federal policy that creates space for girls-only classes as an important direction for integrating gender-equity remedies, schools can take other actions necessary to ensure the success of these classes. The essential training of administrators, teachers, students, and parents about gender equity in general, and girls-only classes in particular, could become part of the culture of schools where such classes are offered. The first and most important step, however, is to empower local practice through broader and more informed federal policy.

REFERENCES

Faludi, Susan. "The Naked Citadel." *The New Yorker* (September 1994): 62–81.

Greene, Maxine. "Sex Equity as a Philosophical Problem." In *Handbook for Achieving Sex Equity through Education*, edited by Susan Klein, pp. 29–43. Baltimore, Md.: The Johns Hopkins University Press, 1985.

Jacklin, Pamela. "The Concept of Sex Equity in Jurisprudence." In *Educational Policy and Management: Sex Differentials*, edited by Patricia Schmuck, W. W. Charters Jr., and R. O. Carlson, pp. 55–72. New York: Academic Press, 1981.

Klein, Susan. *Handbook for Achieving Sex Equity Through Education*. Baltimore, Md.: The Johns Hopkins University Press, 1985.

Sadker, Myra, and David Sadker. *Failing at Fairness: How Our Schools Cheat Girls*. New York: Simon & Schuster, 1994.

Streitmatter, Janice. "An Exploratory Study of Risk-Taking in a Girls-Only Middle School Math Class." *Elementary Education Journal* 98, no. 1 (1997): 15–26.

Stromquist, N. "Sex Equity Legislation in Education: The State as Promoter of Women's Rights." *Review of Educational Research*, 63 (1993): 379–407.

Walkerdine, Valerie. *Schoolgirl Fictions*. London: Verse, 1990.

The Social Context and Women's Learning of Mathematics

17

Mathematics, gender, and equity issues have attracted much research attention in recent decades. Substantial funds and energy have been spent on intervention programs aimed at changing the gender composition of educational and occupational fields in which females are underrepresented. During the 1970s and 1980s, considerable research effort was directed at documenting gender differences in mathematics participation and performance. These comparisons typically assumed that the achievements, experiences, behaviors, and beliefs of males should be accepted as the norm. Measured against these standards, females were frequently found wanting. The removal of perceived barriers and, if necessary, the resocialization of females were seen as the paths to equity.

Over time, the growth of feminist research increasingly challenged these earlier assumptions (Leder, Forgasz, and Solar 1996). The experiences and accomplishments of females, it was argued, should be celebrated as intrinsically worthwhile and of equal value to those of males. Instead of focusing exclusively on females as "the problem," current social structures might need to be changed and popular value positions and norms reevaluated for equity for females to become a reality. The empirical methods and instruments of earlier research began to be scrutinized more carefully, and alternative methods were proposed for exploring the implications of feminist epistemologies for mathematics and its teaching and learning. Women's voices were being heard and acknowledged.

Drawing on two studies that describe experiences of learning mathematics, we focus on the effects of the learning contexts within and beyond the mathematics classroom. The first study was an evaluation of a program of single-sex groupings for mathematics within a coeducational high school. The experiences and perceptions of students enrolled in mathematics at one university were the focus of the second study. Each study could stand alone. In combination, however, powerful images of the effects of the social learning context were evident.

Helen J. Forgasz
Gilah C. Leder
Julianne Lynch

LEARNING MATHEMATICS IN A SINGLE-SEX SETTING

Single-sex groups as the preferred setting for schooling for girls have motivated fierce advocates as well as critics. The case study reported here was carried out in response to a specific request from a coeducational high school in the outer area of Melbourne, Australia. The first appraisal was made in 1993, the year the school introduced single-sex classes for mathematics at the grade 10 level while retaining the coeducational format of all other mathematics classes. (In Australia, mathematics is compulsory until grade 11, although students can select from options with acknowledged different levels of difficulty.)

For many years, a gender imbalance favoring males existed in the enrollment figures in the most demanding mathematics course offered at the grade 11 level. An important aim of the single-sex program was to boost the numbers of girls choosing this particular grade 11 elective mathematics course. The school asked us to monitor the effectiveness of their decision. Interest focused on both the short- and long-term effects of the intervention on females' and males' attitudes and performance in mathematics. Three years later we were invited to reevaluate the program. We were thus able to compare students', parents', and teachers' reactions to the program in 1993 and in 1996.

The Study

In 1993, we gathered data three different times during the school year: early, halfway, and at the end. We used self-report, paper-and-pencil instruments at the first and third data-gathering periods; we conducted interviews halfway through the year. Data were gathered primarily from the students in grade 10 (more than 160 students, equally split between girls and boys), their teachers, and their parents, although some limited information was also obtained from selected students in grades 9 and 11. At each period of data gathering, we asked teachers to rate students' performance in mathematics by using the school's regular assessment program.

In 1996, the time frame for data gathering was restricted to the second semester of the school year. The single-sex program was in a transition phase from grade 10 to grade 9. In 1996, single-sex classes were in place at both of these grade levels so that the grade 10 cohort would not miss out on the experience. Pen-and-paper instruments identical to those used early in 1993 were administered to students in grade 9 ($N = 182$; 77 males [M], 105 females [F]) and grade 10 ($N = 79$; 38M, 41F). The same parent questionnaire was used as in 1993. Interviews were conducted with students in grades 9, 10, and 11, with a few parents whose offspring had been interviewed, and with the mathematics teachers involved in the program.

Results

Although we saw considerable overlap in the responses of the females and the males in 1993, we also identified a number of interesting differences. There were also notable variations in the response patterns of the 1993 and the 1996 cohorts. In this chapter, we emphasize the differences that emerged and draw heavily on the student self-report and interview data and on the self-report data from parents.

Student Voices: Why Single-Sex Classes?

Many of the students we interviewed in 1993 believed that the single-sex classes had been introduced to "make people work harder," to allow "students to concentrate better," and to "stop boys from getting all the attention." Several boys spoke explicitly about girls being distracted or intimidated by boys in other classes and reflected that "girls would work better in a single-sex class." Girls often spoke of having been "too embarrassed before to ask questions," not liking "being made fun of," "being pressured" by boys, and being relieved at getting more work done without the boys' noisy and disruptive behavior. Given the choice, just less than half the boys (48%) said that they would choose a single-sex class again compared with an overwhelming 92 percent of the girls.

In the questionnaire data gathered after the interviews, about 75 percent of the girls, compared with about 25 percent of the boys, indicated that they had enjoyed being in a single-sex class. Very few girls (less than 10%) but more than half the boys had disliked the experience. At the same time, 40 percent of the girls and only about 5 percent of the boys expressed a strong preference for single-sex mathematics classes for the following year. About the same proportion of boys and girls (around 35%) indicated that they would be equally happy in a single-sex or a coeducational setting. Overall, the girls were more positive than the boys about learning mathematics in a single-sex class.

In comparison with the responses in 1993, we saw in 1996 a large decrease in the enjoyment of the single-sex classes among the girls and a smaller decrease among the boys (about 35% and 13%, respectively, had enjoyed the classes). There was also a much smaller proportion of females (about 25%) but a larger proportion of males (about 10%) who wanted single-sex settings to continue into the next grade level.

Did the Single-Sex Classes Make a Difference?

Both at the beginning and at the end of the 1993 school year, teachers rated the performance of the girls in mathematics somewhat higher than that of the boys. Yet the latter consistently rated their performance higher than the girls did and overestimated the rating they thought that they would receive from their teachers. In contrast, the girls underestimated their teachers' assessments. For example, at the end of the year, boys assigned themselves a mean rating of 4.1 out of 5, compared with the girls' mean rating of 3.2. The ratings assigned by the teachers were 3.8 and 4.0, respectively. Thus the single-sex environment did not change the girls' perceptions of their proficiency in mathematics. More promising were the enrollment figures for mathematics in grade 11, the year after the single-sex experience. Unlike in previous years, when relatively few females enrolled in the most rigorous mathematics option, the proportion of females who said that they would choose the subject was approximately the same as for the males. There was a close match between students' mathematics enrollment intentions, expressed at the end of grade 10, and their actual choice in grade 11.

We looked at the enrollment figures for the most demanding mathematics subject offered in grade 12, Specialist Mathematics, and found that in 1995, sixteen boys and only three girls in the school had taken the course. These students were from the 1993 grade 10 cohort that we had studied. We were intrigued that after the 1993 evaluation, when support for the program had been high among the females, the grade 11 enrollment figures the following year (1994) appeared to suggest a measure of success for the intervention program. Yet, the year after, in 1995, very few of the girls had persisted with the most rigorous mathematics offering in grade 12. What had happened to a large number of girls in grade 11 in 1994 that resulted in their not taking Specialist Mathematics in grade 12? The extent to which support for the program had dwindled in the short three years between our two evaluations was also of particular interest. The data gathered from parents suggested at least a partial explanation for the reversal in the patterns of support for the program.

And What about the Parents?

Despite the considerable publicity given to the scheme in 1993 by the school administration—in the school newsletter, through the parents' association, and through direct teacher-parent contacts—more mothers than fathers seemed to know about, and approve of, the introduction of single-sex grade 10 mathematics classes and to have discussed it with their child throughout the year. Well over half the parents (almost 60%) were fully supportive of the scheme; only 7 percent were completely opposed. Those in favor noted that their child "was more relaxed," "was more confident," "seemed to understand mathematics better," and was "getting better marks in math." Those against the scheme argued that "single-sex [boys] classes were more disruptive" or that "schools should mirror the real world."

There were many similarities in the patterns of response among parents in 1993 and in 1996. However, in 1996, fewer parents were aware that the program was in place. Although more parents supported than opposed the program over-all, the level of full support was lower than in 1993. The reduction in support was particularly evident among fathers (50% reduction) and among mothers of daughters (about 15%). Whereas the overall proportion of mothers of sons sup-porting the program remained at about the same level as in 1993 (around 50%), more than 60 percent of the mothers of sons in grade 9 were in that category; interestingly, a smaller proportion of mothers of daughters in grade 9 (about 50%) fully supported the scheme.

Parents were also asked to explain why they believed that the single-sex pro-gram had been introduced and to give reasons for their levels of support for the program. Although the majority of parents perceived that girls were the focus of the program and that they were likely to benefit from it, a fascinating new per-spective on gender issues in mathematics learning, not evident in 1993, emerged in 1996. *Antifeminist* and *backlash* sentiments (the school had gone *too far* in promoting girls as the disadvantaged group) and stronger feelings that boys were the disadvantaged group were expressed by some parents of sons who were often fully supportive of the program. Consider the following comments:

- "In past experience, the boys have been left to manage while the girls are constantly helped." (Mother of son in grade 10, fully supporting the program)
- "As a knee-jerk reaction to the concept of affirmative action." (Father of son in grade 9, unsure of level of support for the program)
- "Math seems to be working better without the girls." (Mother of son in grade 9, partially supportive of the program)

The perception that males are disadvantaged educationally has gained promi-nence in the popular Australian media in recent times. Regarding the outcomes of mathematics education, this argument appears difficult to sustain in general and also at this school. We might infer from a few parents' comments on the questionnaires that outside sources may have shaped their views about single-sex settings. For example, a father of a daughter in grade 9 noted the following:

> Statistics indicate that girls are on average higher achievers than boys. I'm inclined to believe boys are more boisterous and assertive and probably dominate activities in the classroom which probably disadvantages girls. Girls can do without the dis-traction of boys in the room.

Conclusions

The study has not provided unequivocal evidence that single-sex mathemat-ics classes per se address well-documented gender differences favoring males in mathematics-learning outcomes. However, many students, females in partic-ular, enjoyed being in the single-sex classes and believed that they had benefit-ed. Although some students and parents reacted negatively in 1993, actual detri-mental effects appeared to have been minimal. By 1996, the beneficial outcomes of the program were less clear and support for it had waned. The short-term benefits observed in the first instance were not maintained over a longer peri-od. The students' and parents' responses in 1996 reflected a concern now expressed in the wider community that many boys may be disadvantaged by school practices and expectations.

In 1993, a few students had alluded to differences in male and female teach-ers' attitudes toward single-sex classes and to their changed teaching styles in this setting. For example, some boys indicated that their teacher used different language and told more jokes in the all-boys group. In 1996, the comments were more specific. It was clear that in both single-sex and coeducational settings, boys' misbehavior was a problem. In 1996, but not in 1993, several girls complained of too much chatting in the single-sex classes. The parents of a girl in grade 10 claimed that their daughter thought that the girls were taught more slowly than the boys and that the requirements of the assignments they were given were less strict. In an interview, a female teacher who was asked whether she taught her ninth-grade single-sex class differently from coeducational classes said, "Not my manner of planning. I don't think that has been affected. I think I probably spend a lot more time explaining stuff to the girls.... Perhaps my interaction in the classroom is different." A second female teacher's comments suggest that her experiences in single-sex settings had alerted her to differences in the ways in which boys and girls learn mathematics:

> Boys seem to generally follow the really logical approach. Girls ... won't necessarily approach a problem step by step. They might use trial and error or they might draw a diagram of the situation.... Just listening to the kids sometimes when I have explained work on the board and one or two of them do not understand and listening to them explain it to each other, they explain it in different ways than I had thought of, and you can hear the same problem explained ten different ways by ten different girls. It seems to be a more, a wider range of ways of looking at the problem.... Boys seem to be happy with the one method that you do on the board. They just all go into it.

In combination, the comments suggest that in single-sex girls classes, some teachers might adopt different teaching approaches that girls do not always perceive to be to their advantage.

The findings indicate that the efficacy of single-sex groupings—grade level, mandated or voluntary—and the attitudes and beliefs of significant others on students' behaviors warrant further scrutiny.

ADULT LEARNERS OF TERTIARY MATHEMATICS

For the last two decades, efforts have been made to redress gender inequities in society generally. It has been recognized that societal attitudes and expectations have differentially influenced the career aspirations of females and males. Yet, in fields of endeavor previously considered more suitable for males, U.S. and Australian statistics in the mid-1990s still indicate that fewer females than males are enrolled in undergraduate and postgraduate engineering, physical science, computing, and mathematics courses. At the same time, changes in work and lifestyle patterns have led to a large number of mature students engaging in tertiary study. For various reasons, many are choosing to enroll in science and technology courses. The views and perceptions of a group of mature and younger students highlight some of the issues influencing decisions to pursue and persist with tertiary-level mathematics.

The Study

Students enrolled in undergraduate mathematics courses in 1995 at one university in Melbourne, Australia, participated in the study. That year, more than 60 percent of the nearly 16 000 students at the university were female. In mathematics courses, however, the ratio of males to females was approximately 2:1. The faculty of the mathematics department was overwhelmingly male.

Twenty-three students (16M, 7F) agreed to be interviewed in depth. Two (1M, 1F) were enrolled in humanities programs; the others were in science-related courses. Overrepresentative of their enrollment numbers in mathematics courses, a surprisingly large proportion of the interviewees (12/23; 9M, 3F) was at least twenty-one years of age at the commencement of their current tertiary studies.

The students were generally satisfied with their courses. Gender differences in experiences and perceptions were evident, and on some important issues, the views of the younger (Y) and the mature (MA) students differed.

The Tertiary Learning Environment

Telling comments were made about the tertiary learning environment. The life experiences of the mature students appeared to enable them to deal better with the adversities that students might confront. They were more persistent than younger students in seeking and obtaining assistance and seemed to know how to deal with less-cooperative faculty members.

Some females had perceived gender-related and racial discrimination among faculty members and peers. Karen (MA) spoke bluntly:

> Some tutors and lecturers will pay more attention to the women and some don't think women can do math … [and] are more reluctant to help you.… Some of them are racist, arrogant, you know, they don't give much time to some students.

She believed that balance could be found: "If you strike one who is a bit funny towards you, you can find one who is the opposite way." Karen was delighted in having a female teaching in the department. She spoke of the effects on the attitudes of some male students:

> They can see that she does a good job and perhaps even better than some of the other male lecturers.…I think there are a lot of entrenched beliefs [among male students] that women cannot do maths as well as men.

Jan's (Y) experience of sexist behavior was very personal. She claimed that a senior faculty member was a rude "male chauvinist pig": "If you asked him a question he would respond to the male partner whom you were working with." She believed that her learning had suffered, and her helplessness was evident: "You can't do anything about him, that's the way he is, and his belief is that women should not be in math at all."

Perceptions of discrimination based on ethnicity and gender were indirectly confirmed by Jim (MA). When asked whether he believed that any groups were disadvantaged, he said: "Well, I can't really from my perspective, you know what I mean. If I was female or from another country it would be much more easy for me to be aware of such things. But, yes."

Some male students exhibited stereotyped views of females' mathematical abilities. When asked to explain why fewer females than males studied the most demanding mathematics courses at grade 12, three males believed that genetic factors were involved. Sam (Y) said:

> I don't know whether it is a fact or not but women tend to be better off with the humanities type subjects whereas men tend to be better off with the science type subjects and that is just the way their brains work or something.

When we reported the findings of this study to the mathematics department at a seminar, the female lecturer referred to by Karen said that the male students in her classes frequently gave her a hard time. She had been subjected to sexual innuendo and to comments challenging her competence as a mathematics lecturer.

Conclusions

The findings imply that perceptions of mathematics as a male domain persist at the tertiary level, particularly among some male faculty and students.

Discriminatory behavior that may affect some females' capacity to learn may ultimately drive them from persisting with mathematical studies. Students for whom English is not their first language also perceived discrimination that may disadvantage them. For mature students, hardened by the experiences of the workplace, the tertiary learning environment appeared to present less difficulty than for school-leavers.

These two studies reveal the extent to which various social learning contexts can have different impacts on some male and female students. A comment from Karen, a mature student participating in the second study, illustrates the potential interaction of three "learning contexts": single-sex settings, tertiary learning environments, and influences beyond the classroom. Karen had attended a single-sex Catholic girls school populated by students from working-class backgrounds. She said that about "80 percent of [the] girls wanted immediate marriage or secretarial [work].... I think there were only four or five that actually matriculated in the class and … there were thirty or forty." She agreed that she had been unique in being accepted to pursue a dentistry degree.

Karen's perceptions of the "culture" of her single-sex school reveal that simply segregating the sexes is insufficient to challenge gender-stereotyped beliefs about sex roles. As we can infer from the results of the study of single-sex settings for mathematics, the entire school community—students, parents, and teachers—needs to be educated about gender-related issues and to be supportive of such programs to maximize the potential to attain the range of anticipated benefits for female and male students.

The study of tertiary mathematics students also implies that addressing gender issues at the school level alone is insufficient. A different learning environment faces students as they make the transition from school to the tertiary setting. Retaining females in mathematics and related courses remains a challenge. Our findings support Hyde's (1993) contention that tertiary mathematics departments should increase the number of female faculty to serve as role models and to eliminate overt sexism, including sexual harassment. Continued support for mature-aged entry to mainstream tertiary mathematics courses is also needed. Students who recognize the constraints of their earlier educational and career choices, or who experienced limiting personal circumstances, should have the opportunity to fulfill their dreams and realize their potential.

From our studies, we can draw several inferences about the likely impact of influences beyond the classroom, including the print and broadcasting media. Very few of the tertiary students had received career advice about studying mathematics at the higher levels of schooling or at the tertiary level. What then had influenced their decision making? What had contributed to the attitudes toward female students held by the sexist male students and faculty? The parents of the students in single-sex classes were raised in an era when gender issues in mathematics and science-related fields had not found their place on the educational agenda. What had shaped and influenced their beliefs and to what extent had their views influenced their offspring? We believe that the mass media are potent forces interacting with individuals' beliefs and shaping popular opinion within society. Exploiting the media and examining social learning contexts to educate students about the gender-related issues emerging from the studies described in this chapter are the basis of several classroom activities.

IMPLICATIONS OF THE TWO STUDIES

SUGGESTED CLASSROOM ACTIVITIES

Classroom activities aimed at heightening students' critical awareness of the construction of gender and of the factors that can shape attitudes, beliefs, and expectations of males and females and of minority, ethnic, racial, and culturally different groups in society follow. Several of the activities, or variations on them, could also be used with college-level students.

Activity 1: The Print Media

Media reports are an important source of information that affect our views of society. Media coverage of mathematics-related areas has largely been gender-stereotyped. We compared how two groups interpreted selected newspaper items. One group held liberal views on gender and mathematics, that is, did not agree that some jobs are unsuitable for women. The other held traditional or stereotyped views, that is, agreed that girls are more suited to humanities than to mathematics. Both groups read selectively; readers were more likely to focus on information that confirmed their already held beliefs than on information that challenged them.

This phenomenon is implicated in the perpetuation of stereotyped views of gender roles, particularly in mathematics-related areas where so few newspaper articles pose a challenge to stereotyping. The historical link between mathematics and many "male" domains has an impact on students' life choices. A challenge faced by mathematics teachers is to increase students' awareness of the social construction of gender-appropriate roles. The mass media present teachers with rich sources of current, reality-based materials. Newspapers in particular are both accessible and convenient.

The Activity

Challenge students to collect articles about mathematics and mathematics-related career areas. Are women featured? If so, in what capacity? How are they described and represented? Where in the newspaper are women featured most prominently? What messages might this convey? Consider articles in other languages or from other countries. How do they compare?

Activity 2: Equity in Mathematics Textbooks

A new mathematics textbook for seventh-grade students (Stephens et al. 1996) was published in Australia in 1996. It used thirteen cartoon characters to emphasize mathematical ideas and concepts. On inspection, the authors and the publishers appeared to have taken care to include males and females among the cartoon characters and to portray an ethnic mix. Several of the characters had a unisex appearance and were clearly drawn to reflect the interests and appearance of contemporary youngsters—their clothing, hairstyles, and so on.

We were curious about how students would perceive the thirteen characters. Interestingly, a majority of the 235 students surveyed identified six of the characters to be "boys" and seven to be "girls." This result could be interpreted to suggest that gender equity was apparent. However, this was not true. The number of times each character appeared in the book varied greatly. Overall, 56 percent of the total number of occurrences of the characters in the textbook were boys, as identified by the students.

Gender stereotyping was evident in some of the reasons that students gave for their choices of the "sex" of characters. For example, most students (62%) perceived the character shown in figure 17.1 to be a "girl." Gendered identifiers were used as explanations. Several students identified this character as a "girl: because of the lipstick," whereas the skateboard was used to identify the character in figure 17.2 as a boy. Six of the characters were clearly identified as not of Anglo (white) background. The character shown in figure 17.2, for example, was also considered by a majority to be African, Jamaican, or African American.

This study showed how appearance can reinforce gender-stereotyped beliefs. This highlights the importance of gender and ethnic background in the images presented in mathematics textbooks if lingering beliefs that mathematics is the domain of "Anglo males" are to be challenged.

The Activity

Have students investigate their mathematics textbook: word problems, photographs, and illustrations, if applicable. Conduct a frequency count of the females and the males who are featured in the book. In what contexts do females and males appear? What attempts have been made to include women in traditionally "male" contexts? Are traditionally "female" interests depicted? What messages are conveyed? Are there differences between current books and those of earlier years? Similarly, examine the textbook for representations of people from different cultural, ethnic, and racial backgrounds. Is equity evident?

Because college-level mathematics texts are likely to be more traditional and to depict human involvement less frequently than school texts, this may prove a valuable exercise to conduct with college students.

Activity 3: Small-Group Dynamics

Observations of students working in small-group, cooperative-learning settings offer valuable understandings about how children learn mathematics and how cognition and affect interact (e.g., McLeod and Adams 1989; Forgasz and Leder 1996). In one seventh-grade classroom, we observed five students, three girls (Carol, Cheryl, and Jenny) and two boys (Mark and Brian), engaged in an extended problem-based task over eight consecutive lessons. During the monitored period, we found that the two boys were more concerned with the mathematical aspects of the task, whereas the girls worked assiduously on the presentation and writing associated with the report of the findings of the investigation. Like previous researchers, we noted that the division of tasks followed a gender-stereotyped pattern.

Five years later, when the students were in grade 12, we found that only Mark had taken the most rigorous mathematics subject offered at grade 12. The three girls had taken less demanding grade 12 mathematics options. (Brian had changed schools and we were unable to locate him.) We recognize that many factors probably influenced the students' lives and their decisions about what subjects to study at school. Yet, we could not help but reflect on what we had seen in the seventh-grade classroom and muse about whether we could have predicted their future directions.

The Activity

Observe students when they work in pairs or in small-group settings. Do they form same-sex or mixed-sex groups? Do the students allocate roles along gender-stereotyped lines? Elicit the students' views on their choices of work partners and the roles they take in the group task. Discuss your observations with the class.

Activity 4: Equity and Single-Sex Education

Although many people argue that single-sex settings for girls are liberating, there is no guarantee that inequities are overcome by segregation. One recent

Fig. 17.1

Permission to reproduce this character has been obtained from Jacaranda-Wiley.

Fig. 17.2

Permission to reproduce this character has been obtained from Jacaranda-Wiley.

Australian example revealed inequities between two government single-sex schools—one for boys and one for girls (Shorten 1990). A brother and a sister were involved. The girl, like her brother, wanted to learn about computers. The boy's school offered the relevant courses; the girl's school did not. The girl claimed discrimination on the grounds of sex and won the resulting legal case.

A similar situation arose in the United States. A court case resulted when a girl eligible for entry to schools for high achievers was refused enrollment in a single-sex school for boys. She claimed that the school offered better science and mathematics opportunities than a similar single-sex girls school. The case, *Vorcheimer v. School District of Philadelphia*, is described succinctly by Kleinfeld (1995).

Interestingly, both cases involved mathematics- and science-related subject areas.

The Activity

Present the facts of the case described by Kleinfeld (1995) to your class. Organize the students to debate the issues and to decide the outcome of the case. Discuss the actual decisions reached by the courts. Alternatively, discuss whether legal action would be likely if a girls school offered needlework or secretarial studies, but these options were not available in a boys school.

REFERENCES

Forgasz, Helen J., and Gilah C. Leder. "Mathematics Classrooms, Gender and Affect." *Mathematics Education Research Journal* 8, no. 2 (1996):153–73.

Hyde, Janet S. "Gender Difference in Mathematics Ability, Anxiety, and Attitudes: What Do Meta-Analyses Tell Us?" In *The Challenge in Mathematics and Science Education: Psychology's Response*, edited by Louis A. Penner, H. M. Knoff, and D. L. Nelson. Washington, D.C.: American Psychological Association, 1993.

Kleinfeld, Judith S. "The Venerable Tradition of Separate-Sex Schooling." In *Gender Tales: Tensions in the Schools*, edited by Judith S. Kleinfeld and Suzanne Yerain. New York: St. Martin's Press, 1995.

Leder, Gilah C., Helen J. Forgasz, and Claudie Solar. "Research and Intervention Programs in Mathematics Education: A Gendered Issue." In *International Handbook of Mathematics Education, Part 2*, edited by Alan J. Bishop, K. Clements, C. Keitel, J. Kilpatrick, and C. Laborde. Dordrecht, Netherlands: Kluwer, 1996.

McLeod, Douglas B., and Vera M. Adams, eds. *Affect and Mathematical Problem Solving: A New Perspective*. New York: Springer-Verlag, 1989.

Shorten, Anne. "Equality of Educational Opportunity in Australia." *Education Law Journal* 2 (1990): 265–97.

Stephens, Robert, B. Woods, G. Sotirou, J. Seymour, G. O'Loghlen, B. Simons, C. Devlyn, V. Sotirou, J. Dolman, and H. Neyland. *Maths Power for the Curriculum and Standards Framework—Victoria*. Milton, Queensland, Australia: Jacaranda Press, 1996.

Calculate the Possibilities

A Program in Mathematics and Science for Young Women

18

Bernadette H. Perham
Rebecca L. Pierce

Calculate the Possibilities was a four-week summer residential program for Indiana high school women who had completed grade 10 or grade 11. It can serve as one model of the types of partnerships that high schools and colleges and universities can design to encourage young women to consider mathematics- and science-based careers. The program had a dual focus: career awareness and skill development in the areas of science, engineering, and mathematics (SEM). Specifically, the major goals of the program were to increase the awareness of young women about careers in SEM; build their mathematical confidence and competence; have them interact with women who have SEM careers or are preparing for such careers; and, in general, empower the young women.

RECRUITING AND SELECTING PARTICIPANTS

The project did not specifically target girls who had already selected a major in mathematics, science, or engineering. It also included college-capable students whose choice of career most likely could be influenced by this experience.

Students selected for this program had completed either their sophomore or junior year in high school by the beginning of the residential program. Applicants were required to have completed the following courses: algebra; geometry; and at least one laboratory science course—biology, chemistry, or physics. To be admitted into the program, applicants needed a grade-point average (GPA) of at least 3.0 (A = 4.0) in mathematics and science courses and an overall GPA of at least 2.75.

Each application packet included a student form; a school nomination form; a resource-teacher commitment form; and two forms for letters of recommendation, one from a guidance counselor and one from a mathematics or science teacher. Each applicant also wrote an essay of approximately 500 words that addressed her experiences with mathematics and the sciences and described her career intentions and aspirations.

Each applicant identified a home school mathematics or science teacher willing to support the follow-up research project and to accompany her to the Saturday conference during the following fall semester. Each resource teacher received a $180 stipend for participating in the follow-up activities.

Bernadette Perham died in 1996.

THE NATURE OF THE PROGRAM

Through career seminars, company site visits, and panel discussions, the participating young women learned about different SEM careers and the educational background they require. The young women took planned tours of major Indiana companies where they had the opportunity to interact with professionals in pharmaceutics, medical care, naval avionics, market research, engineering, and quality control. The students summarized their visits in a career portfolio.

In addition, a university career consultant was responsible for assisting the young women in exploring their career objectives and goals. Each young woman completed a career assessment battery that examined her skills, interests, and abilities and received a list of occupations that matched her career interests. They also learned to use a computerized guidance-information system to gain access to a data bank of information about occupations and academic courses and programs.

An important goal of the program was building the young women's mathematical competence and confidence. In the context of creative problem solving within a variety of SEM areas, the young women acquired skill in using a spreadsheet (Excel), a word processor, statistical software, and a graphing calculator (TI-92). They were helped in the computer laboratory by female undergraduate and graduate assistants who demonstrated their comfort and competence with mathematics and technology.

During the first week, these sessions introduced the young women to various technologies, including the World Wide Web, e-mail, chat groups, ERIC, library search software, and the TI-92 graphing calculator. During the second week, the young women used Microsoft Excel to explore Heron's formula, graph three-dimensional surfaces, and solve systems of equations by using matrices. Although each young woman had access to a computer, guided discovery in collaborative groups was the mode of instruction. In the third and fourth weeks, the young women focused on using the many capabilities of the TI-92 calculator to solve problems that involved applications in geometry, algebra, and statistics. Modes of instruction included individual and group work, as well as hands-on activities.

At several points in the program, the young women met, talked with, and interacted with women who already had SEM careers or were studying to work in mathematics- and science-related fields. During field trips, the students met women who were successfully employed in such careers and were able to talk with them about the preparation that their careers required and the challenges and rewards of such careers. In the technology sessions, young women who were students at Ball State University and were pursuing SEM majors and careers worked with the high school women as they learned new skills. Here the young women learned from capable and mathematically, scientifically, and technologically competent women who were close in age to themselves.

In addition to these interactions, the young women from the high schools worked on research projects with mentors from the university and with their high school resource teachers. These experiences allowed additional interactions with women who had SEM careers.

As part of the residential program, the young women participated in weekly group recreational activities, such as Jazzercize, bowling, softball, swimming, and hiking. All activities of the program were designed to empower the young women and to make them feel strong and competent.

The Research Component of the Program

Calculate the Possibilities afforded opportunities for the young women to acquire skills in using statistics and technology to support future research efforts. In collaboration with a SEM mentor from Ball State University, each young woman gained hands-on research experience in a university laboratory setting. Each also independently solved a research problem at her home school with the support of the on-site resource teacher and her university mentor.

Working in small groups led by a SEM mentor, the young women used the scientific method and various laboratory techniques. The SEM mentors introduced the young women to the library as a resource for research and showed them how to conduct a literature search for writing scientific reports and research proposals within the context of their collaborative research projects. The young women selected projects in biology, chemistry, nutrition, physics, or psychological science. The small-group activities for each area were as follows:

Biology: Assessing the ability of new compounds to affect the immune system and various microbes. This activity involved performing several modern laboratory techniques concerning DNA and proteins. The students used the TI-92 calculator to do numerical calculations and graphics.

Chemistry: Investigating the use of a new thin-layer chromatography technique for separating organic compounds. The students used infrared microspectroscopy to detect and identify the compounds *in situ*. This group used Excel to perform the analyses, both numerical and graphical, on the collected data.

Nutrition: Analyzing the degree of correlation between fat and calcium intake among college students. Approximately fifty female university students, who had not taken prior nutrition classes, participated in the study. The high schoolers collected three-day diet food records before and after nutrition education and analyzed them by using the Nutritionist IV computer software.

Physics: Studying basic electronic circuits and modern semiconductor devices. Specifically, the students became familiar with DC and AC circuits, multimeters, oscilloscopes, and semiconductor components. This group also used the TI-92 calculator to support their work.

Psychological Science: Investigating how women and men perceive their gender roles. The high school women used state-of-the-art computer technology to collect and analyze data obtained from the subject pool of the university's psychology department.

Each research group developed the mathematics necessary to understand the problem being investigated. Although mostly informal in nature, the mathematics was embedded in the activities in the laboratory setting. However, in analyzing results, each group used newly acquired graphing and statistical skills. Each young woman realized, as shown by comments and written evaluations, that a background in mathematics is essential to pursuing advanced study in most subject areas.

At the conclusion of the four-week program, each of the five research groups gave an oral presentation of its research results to the other participants, the codirectors of the program, and the SEM mentors. During the residential part of the program, each student identified a research project to carry out independently at her home school with the support of her university mentor and home school resource teacher.

The Academic-Year Component of the Program

During the following fall, the participants and their home school resource teachers spent a day on the university campus to meet with the program staff. At that time, the resource teachers were given a packet of materials and received the training needed to support the research efforts of their respective students. The resource teachers became acquainted with the research techniques and technology used by the young women in the summer program.

Participants served as peer role models in their home schools as they solved their individual research problems, orally presented their written results, and shared the information in their career portfolios.

Participants left the residential program with information on SEM careers, technological skills that support SEM problem solving, and experience in using the scientific method of discovery. They explored SEM careers and interacted with SEM professionals. They were introduced to spreadsheets, software packages, graphing calculators, word processing, and Internet communications. They engaged in group and independent research projects supported by SEM mentors and high school resource teachers. An added benefit of the Calculate the Possibilities program were the twenty-four SEM ambassadors in schools around the state who could significantly effect awareness of, and interest in, SEM careers of the young women in their schools.

What Can Be Done to Increase Enrollments of Talented Female Mathematics Students?

A Study of the Impact of Middle School Intervention Programs

19

Harvey B. Keynes
Andrea Olson
Karen Singer Cohen
Tracy Bibelnieks

The University of Minnesota Talented Youth Mathematics Program (UMTYMP or the Program) is a statewide program aimed at providing an alternative educational experience for Minnesota's most mathematically talented students. Students in grades 5–8 who have scored above the 95th percentile on any standardized mathematics test are identified and invited to take part in a qualifying examination. From approximately 1400 students who take the test annually, 140 to 150 are invited to participate.

The Program offers an intense academic environment and a culture of mathematics through a sequence of uniquely designed accelerated mathematics courses. Classes are held once each week for two hours after school for thirty weeks throughout the academic year. The students average eight to ten hours of homework weekly. In the high school component, students study a customized sequence in algebra (equivalent to first-year algebra and second-year algebra) in the first year, and geometry and mathematical analysis in the second year, for which they earn a total of four high school mathematics credits. The high school component is taught by outstanding certified high school mathematics teachers and undergraduate teaching assistants, many of whom graduated from the Program.

Following the high school component, the students enter the college component: specially created university mathematics courses that are taught by mathematics faculty using a conceptually based reform approach and problem-based curricula. Students study differential and integral calculus in the first year,

The intervention programs and this study were supported in part by grants from the Bush Foundation (Saint Paul, Minnesota), the National Science Foundation, and the Department of Energy. The opinions expressed in this paper are those of the authors and do not reflect in any way policies or opinions of the supporting agencies.

We would like to thank the many students, staff, and faculty who are part of our current program and whose cooperation and support significantly contributed to this study.

linear algebra and differential equations in the second year, and linear analysis—multivariable calculus using linear algebra and geometry—in the third year. For students who are still in high school during the fourth year, a topics course in advanced undergraduate mathematics is offered. Since students graduate from high school at varying points in the Program, the calculus courses are designed to allow the students to move easily into subsequent mathematics courses at leading universities. A student earns eight to ten University of Minnesota mathematics credits for each successfully completed year of the UMTYMP college component.

THE INTERVENTIONS

The interventions were designed on the basis of a sampling of best practices from other programs. They were supplemented by other activities at the University of Minnesota, including the National Science Foundation–sponsored Young Scholars Program's Early Alert Initiative and Project YES and the Department of Energy's Pre-Freshman Enrichment Program (PREP). The primary objectives of these interventions were to recruit and retain students from underrepresented groups in mathematics and science. The interventions, based on the following broad set of objectives, were designed to provide curricula and experiences to motivate these students to study mathematics:

- To provide a comfortable, interesting setting in which students could learn and enjoy mathematics while becoming more aware of its applications and its role in society
- To offer opportunities for students to become familiar with mathematical problem solving in a stimulating, small-group environment
- To introduce students to innovative mathematics and research methodology
- To create an environment that facilitates one-on-one interaction between students and successful faculty, business people, and researchers
- To help students become aware of communication problems in mathematics and to improve their communication skills
- To increase understanding of career and academic opportunities in mathematics and related areas (e.g., engineering, chemistry, and physics)
- To address social and cultural issues facing talented female students, students of color, and economically disadvantaged students
- To enhance UMTYMP dissemination—highlighting the value of mathematics programs that provide enrichment in mathematics and its applications while preparing and encouraging students to test and participate in the Program
- To identify a larger number of target students

A series of academic-year events culminating in a three-week summer enrichment institute were developed to meet these objectives. The monthly activities, held for three to four hours on a weekday late afternoon or early evening or on a Saturday morning, focus on mathematics topics outside the standard school curriculum, such as symmetry and the fourth dimension, and on opportunities in mathematics. Several events include activities for family members of students who come close to qualifying for ther program and current UMTYMP students who provide support, motivation, and mentoring for future Program participation. The family activities include career panels, computer lab presentations, and mathematics fairs. In the fall, students and their parents are encouraged to attend the orientation, which introduces the intervention program, its expectations and goals, and the types of mathematics to be covered.

Subsequent activities emphasize small-group instructional workshops, research-oriented interactive demonstrations, career exploration, faculty interactions, tours and demonstrations, informal mentoring, and question periods. For example, during a unit on computerized tomography (CT) scans, students were introduced to the problems of reconstructing information and graph theory problems. They used linear equations to model a CT scan reconstruction problem and discussed how to interpret the solutions. The unit concluded with a visit to the radiology department of a hospital to see CT scanners in use and to hear a presentation by a staff physicist (Keynes, Olson, and Singer 1997). (For additional information about the intervention-program activities, see www.math.umn.edu/itcep/.)

Each monthly event consists of a directed group activity, a demonstration or tour, and time for socialization. For example, one of the most successful events was the graphing calculator workshop, which comprised the following components:

- Interactive demonstrations to small groups (15 to 20 students) on the basic use and functions of the graphing calculator, led by an instructor and teaching assistant
- Rotation of teams through three stations at which they used graphing calculators with Calculator-Based Laboratory systems (CBLs) to graph data from experiments that explored motion, sound, and temperature
- Reconvening of the groups to discuss their experiences and the implications of their findings.
- A refreshment break for socialization prior to or after the rotations
- A short (four to five question) evaluation of the event that the participants complete as the concluding activity

Students who attended at least six academic-year events are invited to participate in the nonresidential summer enrichment institute. The institute emphasizes mathematics and related topics as a culture and a method of scientific inquiry, mathematics as a career enabler, and mathematical thinking as a challenging but exciting opportunity for each participant.

The summer institute curriculum has some special features:

- Close interactions with research mathematics faculty and graduate and undergraduate students
- Innovative content material and instructional approaches that emphasize small-group problem solving and group interactions
- Hands-on computer and calculator activities held at the University Institute of Technology labs and centers

The mathematics activities in the summer are similar in format to those during the academic year, but they have more depth and longer-term projects. Institute curricula have included units on chaos and computing, the physics of motion, and architecture and civil engineering, along with forays into the applications of mathematics to such areas as the health sciences and aerospace engineering. For example, a five-day unit on structures and bridge design comprised a class discussion on the mathematics of structure, a presentation on bridge building by a civil engineer, a visit to a laboratory that studies the integrity of building materials, and a field trip to local bridges. The unit concluded with the students constructing their own wooden bridge models and entering them in a competition on bridge strength, which was held during the program's open house.

THE STUDY

This study examines female students in the Twin Cities component of UMTYMP. It compares those who participated in the intervention programs starting in the years 1989–1994 with their peers who did not. In the broadest sense, *intervention females* (IF) refers to the female students who were near qualifiers (within 10 points of qualifying on the entrance exam prior to the fall of the academic year); participated in at least one "pre-UMTYMP" mathematics enrichment event, 1989–1995 (specific components in the study restrict the 1992–1995 IF groups to only those students who participated in three or more events or the summer institute); and then were retested at the early spring exam. Female students who did not participate in any intervention events are referred to as *non-intervention females* (NF), whether they were near qualifiers or qualified the first time they tested. Data on the NF students who were near qualifiers and retested 1993 to 1995 were studied as a subset to accurately reflect the influence of intervention on retesting and enrollment. Specific components in the study restrict the 1992 to 1995 NF group to only those female students who participated in two or fewer events and did not attend the summer institute.

Student achievement in UMTYMP was studied from two perspectives—grades received and persistence through the program. These data were analyzed for all UMTYMP female students who enrolled in UMTYMP first-year algebra in fall 1990 through 1995. As a supplement, UMTYMP teachers in fall 1995 assessed each current student's individual achievement on the basis of effort and commitment, quality of class participation, comprehension of the subject material, achievement on homework and exams, and overall success in the program. In addition, 1995–96 UMTYMP students were asked to respond to surveys that reflected their behavior as Program participants indicating faculty and student and peer collaboration.

Retesting and Enrollment in UMTYMP

Each spring since 1990, the 50-point UMTYMP qualifying exam has been offered on a special day to female and other underrepresented populations in mathematics who nearly qualified for the program in the previous year; among these students are the intervention females (IF). The test-score gains, qualification results, and entrance statistics for the IF group are given in table 19.1 for 1993–95. The 1990–92 groups exhibited similarly strong levels of qualification and enrollment in UMTYMP but are not included here because intervention attendance records are not as clear for this period.

Table 19.1 shows the impact that the intervention program has had on the enrollment of females in UMTYMP, contributing 46 percent, 38 percent, and 59 percent of the entering female UMTYMP population for 1993, 1994, and 1995, respectively. The data also reveal that the average score gain of qualifiers is higher for the IF students than for the NF students who retested concurrently without having participated in the intervention events. This result suggests that there is a benefit to providing an intervention that makes it inviting for students to take the test a second time.

Student Achievement and Retention

Once the IF students were enrolled in the high school component of the Program, achievement data indicate very high levels of success for these students (table 19.2).

The IF student performance at the beginning of the college component is somewhat lower, however; only 39 percent of the IF group achieved at the B+ or higher level in comparison with 59 percent of the NF group for the first

Table 19.1
Retest Scores and UMTYMP Entrance Data

Intervention Females (IF) Retesters, 1993–1995

Retest Year	No. Retesting	No. Qualified	Average Score Gain of Qualifiers	No. Entering in Fall	Percent of Entering Students from This Group	Percent of Entering Females from This Group
Spring 1993	40	22	7.82	21	16	46
Spring 1994	55	23	6.13	20	16	38
Spring 1995	44	25	7.44	22	16	59

Nonintervention Females (NF) Retesters, 1993–1995

Retest Year	No. Retesting	No. Qualified	Average Score Gain of Qualifiers	No. Entering in Fall	Percent of Entering Students from This Group	Percent of Entering Females from This Group
Spring 1993	22	8	6.75	7	5	15
Spring 1994	23	4	5.50	4	3	8
Spring 1995	19	7	7.14	3	2	8

Table 19.2
Fall 1995 Achievement Data for IF and NF Students

High School Component Courses	Percent of Students Achieving a B+ or Greater	
	IF Students	NF Students
First-year algebra	72	56
Second-year algebra	69	59
Geometry	68	71
Math Analysis	61	40
Calculus I (Sem. 1)	39	59

semester of Calculus I, fall 1995. One possible contributing factor may be that the IF students struggle more with the subject material than the NF students do. The transition from high school mathematics (first-year and second-year algebra and geometry and math analysis) to the more conceptual nature of the material in the UMTYMP calculus courses and the expectations of an honors-level college sequence may present significantly new intellectual hurdles for these students. Researchers also observed that students' willingness to seek assistance with this level of material seemed to need reinforcement among the IF students. In response to a survey given to the first-semester calculus students, only 25 percent of the IF students reported contacting either the teacher or the teaching assistant at least half the time when they needed assistance, contrasted with 56 percent of the NF population.

Teacher assessment information was reported as aggregate data over all the high school component courses and correlates well with the findings on grade distributions. Test achievement, quality of participation in class discussions, homework achievement, and commitment to the program were not significantly different between the IF and the NF groups. On comprehension of the subject material, 79 percent of the IF group were rated very good to excellent, in comparison with 66 percent of the NF group. For contributions to cooperative learning during class sessions, 54 percent of the IF group were rated very good to excellent compared with 34 percent of the NF group. For overall success in the program, teachers rated 81 percent of the IF group as very good to excellent in comparison to 71 percent of the NF group.

Teacher assessment for the calculus component also reflected a decline in performance, with significantly fewer of the IF students ranked as very good to excellent for academic achievement. In balance, it is worth noting that the teacher assessments of effort and commitment to the Program and contributions to cooperative-learning activities were not significantly different between the two groups.

Some of the most positive results are in the retention data for the IF and the NF groups through the second semester of Calculus I (table 19.3).

Table 19.3
Retention Data for IF and NF Students, 1992–96

Retention from First-Year Algebra through	Percent of IF Students Retained	Percent of NF Students Retained	
Second-year algebra	97	90	(*Critical Point 1)
Geometry	75	65	
Math Analysis	70	63	
Calculus I-Sem. 1	45	33	(**Critical Point 2)
Calculus I-Sem. 2	33	29	

*Critical Point 1 indicates the transition from the second-year algebra curriculum to the more analytical and visual geometric content.

**Critical Point 2 indicates the increased conceptual nature and expectations of the honors-level college calculus curriculum.

The analysis of the data indicates a strong commitment to remain in the Program among the IF students. This finding is particularly noticeable at the first critical point during the transition from traditional coursework to challenges of first- and second-year algebra, and it is reinforced at the second critical point—the completion of the first semester of Calculus I. As illustrated in table 19.3, only 4 of 126 IF students taking first-year algebra withdrew during the course (3%), in comparison to 16 of 155 NF students (10%). Beyond first-year algebra through the end of the first year in the college component, persistence for the IF group continues to remain higher than for the NF group. A particularly important observation is that despite a decline in achievement for IF students versus NF students, the IF students show a stronger tendency to persist through the first semester of Calculus I and into the second semester. Of the IF students who entered the first-year algebra level, 45 percent successfully completed the first semester of Calculus I in comparison to 33 percent of NF students. Beyond Calculus I, retention data may have questionable reliability because the numbers of students are small and easily affected by events such as graduation from high school and families moving away from the area.

The additional data collected on student behavior indicated that the IF students formed strong friendships in the intervention programs and used peer collaboration in UMTYMP more often than the NF students. These habits persisted into the calculus component and are also reflected in the willingness of the IF students to ask more questions in class. No other aspects appeared to be significantly different between the two groups.

CONCLUSIONS AND SUMMARY

The overall purpose of this study was to determine whether the Program intervention activities were effective in attracting females who enjoy and are successful in a challenging mathematics program such as UMTYMP and to identify areas for further improvements. The results indicate that these interventions help increase the number of successful female students in the program and point out potential needs for additional interventions.

The female intervention group significantly determines the overall female population of UMTYMP. Frrom 1993 through 1995, 47 percent of the females entering first-year algebra came from the IF pool. The IF students have significantly higher prospects of qualifying for UMTYMP on retesting and a slightly higher average gain than the NF group.

Although the IF students might be viewed as initially less mathematically at ease than the overall UMTYMP population, any differences in academic performance or behavior in the high school component seem to be slight. Moreover, the IF students appear to assimilate extremely well into the overall UMTYMP student population when they enter the Program. This point is underscored by teacher assessments showing high school component IF students to be stronger in comprehension and overall to be more successful with the curriculum than NF students in the early years. They are also more willing to contribute in cooperative-learning activities. In terms of grade achievements, the IF group appears to do at least as well as, and possibly slightly better than, the NF group throughout the algebra, geometry, and mathematical analysis classes. Retention rates all the way through the program seem to be stronger for the IF group, especially at the two critical points—the transition from traditional coursework to the challenges of first- and second-year algebra and the successful completion of Calculus I.

One point at which there is a slight decrease in academic achievement is the beginning calculus level. Grades and teacher assessments seem to indicate that more IF students than NF students find the material challenging. Although the IF students found faculty help unnecessary for the high school component, the culture of seeking help from faculty and teaching assistants seems to need reinforcement at the calculus level. This suggests that implementing an academic intervention involving staff in support of the IF population in the beginning calculus classes would be useful. Since the IF group in calculus still has a strong commitment to the program and higher retention rates despite a lower performance level, these interventions would probably have an immediate impact in overall performance. Anecdotally, some of the strongest students in the college component classes come from the IF group. This is particularly striking when we realize that they were accepted into the program on the second try.

In summary, the intervention activities have created a positive environment for females in UMTYMP. The key to having female students enjoy UMTYMP, succeed academically, and persist in the Program is to get them to enroll in the first place. The intervention program has been successful in increasing female enrollment. Moreover, these interventions appear to be very robust for the target female population. The success of the intervention program for female students persists even with the inclusion of males in the enrichment programs and in the retesting pool. In fact, when male students were included as a significant part of the intervention population in 1994–95, the percentage of male students from the intervention programs who qualified to enter UMTYMP on a retest did not seem to be as large as that of females.

Several studies have claimed that scores for talented females on mathematics aptitude and achievement tests are lower than for their male counterparts. A study reported in the *Chronicle of Higher Education* (Gwaltney 1995) found that on several tests, male students scored in the top 5 percent seven times more frequently than females. Our study indicates that UMTYMP and its intervention programs create more balanced opportunities for both male and female students to become mathematically proficient at the highest levels. These interventions could serve as models or guides for other programs in mathematics that wish to increase the participation and success of talented female students.

BIBLIOGRAPHY

Gwaltney, Corbin, ed. "Footnotes." *Chronicle of Higher Education*, 14 July 1995, p. A6.

Keynes, Harvey, Andrea Olson, and Karen Singer. "Developing Mathematics Enrichment Workshops for Middle School Students: Philosophy and Sample Workshops." *Journal of Mathematics and Science: Collaborative Explorations* [Virginia Mathematics and Science Coalition: National Alliance of State Science and Mathematics Coalitions] 1, no. 1 (Fall 1997).

Keynes, Harvey B. "Can Equity Thrive in a Culture of Mathematical Excellence?" In *New Directions for Equity in Mathematics Education*, edited by Walter G. Secada, Elizabeth Fennema, and Lisa Byrd Adajian, pp. 57–92. New York: Cambridge University Press, 1995.